Andreas Schwen

Agricultural Impacts on Soil Hydraulic Properties

Andreas Schwen

Agricultural Impacts on Soil Hydraulic Properties

Measurements and Simulations

Südwestdeutscher Verlag für Hochschulschriften

Impressum/Imprint (nur für Deutschland/only for Germany)
Bibliografische Information der Deutschen Nationalbibliothek: Die Deutsche Nationalbibliothek verzeichnet diese Publikation in der Deutschen Nationalbibliografie; detaillierte bibliografische Daten sind im Internet über http://dnb.d-nb.de abrufbar.
Alle in diesem Buch genannten Marken und Produktnamen unterliegen warenzeichen-, marken- oder patentrechtlichem Schutz bzw. sind Warenzeichen oder eingetragene Warenzeichen der jeweiligen Inhaber. Die Wiedergabe von Marken, Produktnamen, Gebrauchsnamen, Handelsnamen, Warenbezeichnungen u.s.w. in diesem Werk berechtigt auch ohne besondere Kennzeichnung nicht zu der Annahme, dass solche Namen im Sinne der Warenzeichen- und Markenschutzgesetzgebung als frei zu betrachten wären und daher von jedermann benutzt werden dürften.

Coverbild: www.ingimage.com

Verlag: Südwestdeutscher Verlag für Hochschulschriften GmbH & Co. KG
Heinrich-Böcking-Str. 6-8, 66121 Saarbrücken, Deutschland
Telefon +49 681 37 20 271-1, Telefax +49 681 37 20 271-0
Email: info@svh-verlag.de

Approved by: Wien (Österreich), Universität für Bodenkultur, Dissertation, 2011

Herstellung in Deutschland:
Schaltungsdienst Lange o.H.G., Berlin
Books on Demand GmbH, Norderstedt
Reha GmbH, Saarbrücken
Amazon Distribution GmbH, Leipzig
ISBN: 978-3-8381-3113-9

Imprint (only for USA, GB)
Bibliographic information published by the Deutsche Nationalbibliothek: The Deutsche Nationalbibliothek lists this publication in the Deutsche Nationalbibliografie; detailed bibliographic data are available in the Internet at http://dnb.d-nb.de.
Any brand names and product names mentioned in this book are subject to trademark, brand or patent protection and are trademarks or registered trademarks of their respective holders. The use of brand names, product names, common names, trade names, product descriptions etc. even without a particular marking in this works is in no way to be construed to mean that such names may be regarded as unrestricted in respect of trademark and brand protection legislation and could thus be used by anyone.

Cover image: www.ingimage.com

Publisher: Südwestdeutscher Verlag für Hochschulschriften GmbH & Co. KG
Heinrich-Böcking-Str. 6-8, 66121 Saarbrücken, Germany
Phone +49 681 37 20 271-1, Fax +49 681 37 20 271-0
Email: info@svh-verlag.de

Printed in the U.S.A.
Printed in the U.K. by (see last page)
ISBN: 978-3-8381-3113-9

Copyright © 2012 by the author and Südwestdeutscher Verlag für Hochschulschriften GmbH & Co. KG and licensors
All rights reserved. Saarbrücken 2012

Acknowledgement

Scientific research is the result of the efforts not only of one scientist, but of a team. Therefore, I like to thank all the people that contributed to the success of this thesis.
First of all, I would like to thank Prof. Willibald Loiskandl from the Institute of Hydraulics and Rural Water Management at the University of Natural Resources and Life Sciences Vienna (BOKU) for supervising me during my doctoral research. He gave me encouragement and a strong degree of freedom that allowed me to find my specific field of research interest. Prof. Loiskandl supported my intentions for research visits abroad and attendance to international scientific conferences. In his function as head of the institute, he gave me the possibility to gain teaching experience and participate in the administration of the university.
Thanks for scientific inspiration are mainly due to Dr. Gernot Bodner from the Institute of Agronomy and Plant Breeding at BOKU University. During our field work and extensive discussions and calculations, he generously offered scientific ideas that led to the presented thesis. In this context, I thank Peter Scholl, who helped with the field measurements in Raasdorf.
One key stone for the success of this thesis was a research visit at the Lincoln University in Christchurch, New Zealand, in 2010. During this time, I was supervised by Prof. Graeme Buchan. I would also like to adress thanks to Sam Carrick from Landcare Research, Lincoln, and the team from Plant & Food Research Lincoln, by name Guillermo Hernandez-Ramirez, Erin Lawrence-Smith, Sarah Sinton, and Mike Beare. During my research visit in New Zealand, I was extensively inspired due to a scientific exchange with Markus Deurer and Brent Clothier from Plant & Food Research, Palmerston North.
I also acknowledge scientific inspiration by Kai Schwärzel from the Technical University of Dresden, Jirka Šimůnek from the University of California, Riverside, and Donald R. Nielsen from the University of California, Davis. Since this cumulative thesis consists of three peer-reviewed publications, I gratefully acknowledge the anonymous reviewers for their helpful and constructive comments.

Content

1. Introduction 1
2. Hypotheses and objectives 2
3. Structure of the study...2
4. Dissemination 3
5. Hydraulic Properties and the Water-Conducting Porosity as Affected by Subsurface Compaction using Tension Infiltrometers 5
   5.1   Introduction...6
   5.2   Materials and Methods 7
       5.2.1   Experimental Site..7
       5.2.2   Tension Infiltration Measurements 8
       5.2.3   Wooding's Equation 9
       5.2.4   Hydraulically Effective Porosities and Flow-Weighted Mean Pore Radius 10
       5.2.5   Inverse Parameter Estimation and Calculation of Treatment Means 12
       5.2.6   Statistical Analysis..13
   5.3   Results and Discussion 15
       5.3.1   Soil Compaction and Infiltration Measurements 15
       5.3.2   Near-Saturated Hydraulic Conductivity 15
       5.3.3   Hydraulically Effective Porosity...17
       5.3.4   Inverse Parameter Estimation 22
   5.4   Conclusion 23
   5.5   References...24
6. Temporal dynamics of soil hydraulic properties and the water-conducting porosity under different tillage 29
   6.1   Introduction...30
   6.2   Materials and Methods 32
       6.2.1   Experimental Site..32
       6.2.2   Tension Infiltrometer Measurements..32
       6.2.3   Data Analysis Procedure 33
       6.2.4   Near-Saturated Hydraulic Conductivity 33
       6.2.5   Inverse Estimation of Soil Hydraulic Parameters.....................................34
       6.2.6   Description of Water-Conducting Porosities 35

	6.2.7	Statistics	37
6.3		Results and Discussion	39
	6.3.1	Soil Physical Properties	39
	6.3.2	Infiltration Measurements and Near-Saturated Hydraulic Conductivity	39
	6.3.3	Soil Hydraulic Properties	40
	6.3.4	Water-Conducting Pore Characteristics	46
	6.3.5	Temporal vs. Management-Induced Dynamics of Hydraulic Properties	48
6.4		Conclusion	49
6.5		References	51

7. Time-variable soil hydraulic properties in near-surface soil water simulations for different tillage methods 57

7.1		Introduction	58
7.2		Materials and Methods	59
	7.2.1	Experimental Site	59
	7.2.2	Sampling and Infiltration Measurements	61
	7.2.3	Near-Saturated Hydraulic Conductivity	62
	7.2.4	Inverse Estimation of Soil Hydraulic Parameters	63
	7.2.5	Simulation Model	64
7.3		Results and Discussion	66
	7.3.1	Climatic conditions and Soil Water Content	66
	7.3.2	Temporal Dynamics of Soil Hydraulic Properties	69
	7.3.3	Performance of Near-Surface Water Simulations	69
	7.3.4	Water Balance Simulation	72
7.4		Conclusion	73
7.5		References	74

8. Final conclusion 79
9. References 81
10. Index of tables 82
11. Index of figures 83

Abstract

Agriculture affects the soil's capability for water infiltration and storage. In the present thesis the agricultural impacts in terms of subsurface compaction and different tillage techniques on the soil hydraulic properties and the underlying pore characteristics were analyzed. In-situ tension infiltrometer measurements were used to quantify changes in the near-saturated hydraulic conductivity, the water-conducting porosity, and inversely estimated parameters of the van Genuchten/Mualem (VGM) water retention model. In the first part of this thesis, infiltration measurements in differently compacted subsoil treatments of a silt loam soil in Lincoln/New Zealand were used to characterize the effects on the soil's porosity and its associated water-conducting properties. A high susceptibility to the applied compaction was found, as the saturated hydraulic conductivity of the heavy compacted soil was 81% less than that of the loosened soil. This soil property may be used as a proxy for compaction-induced changes in hydraulic characteristics. Increasing compaction decreased the number of hydraulically effective macropores, reduced the flow-weighted mean pore radius, and the α_{VG} parameter of the VGM model. In the second part, the impact of different tillage techniques – conventional (CT), reduced (RT), and no-tillage (NT) – and their temporal dynamics on another silt loam soil in Raasdorf/Austria were captured using repeated tension infiltrometer measurements. The results show that the near-saturated hydraulic conductivity was in the order CT > RT > NT, with larger treatment-induced differences in the mesopore range. The VGM model parameter α_{VG} was in the order CT < RT < NT, with high temporal variations under CT and RT. NT resulted in the greatest water-conducting pore radii. The results give indirect evidence that NT leads to a greater connectivity and smaller tortuosity of macropores, and to a better temporal stability. Variations in mesopore-related quantities could be explained sufficiently by an interaction of tillage and time. In the third part, the inversely estimated hydraulic parameters were used to parameterize a soil water simulation for two consecutive seasons. Simulated water dynamics of the near-surface soil with constant and time-variable hydraulic parameters were compared to measured water contents. The use of time-variable hydraulic parameters significantly improved simulation performance for all treatments, resulting in average relative errors below 13 %. The study demonstrates the applicability of inversely estimated hydraulic properties for soil water simulations. The simulated water balance indicated that RT and NT result in a better near-surface water storage than CT. This may increase water efficiency, especially under dryer climatic conditions.

Key words: soil hydraulic properties, hydraulic conductivity, pore-size distribution, tension infiltrometer, compaction, tillage, temporal variability

Kurzfassung

Landwirtschaftliche Nutzungen beeinflussen die Fähigkeit von Böden, Wasser zu infiltrieren und zu speichern. In der vorliegenden Doktorarbeit wurden Bewirtschaftungseinflüsse im Sinne von Unterbodenverdichtung und unterschiedlicher Bodenbearbeitungstechniken auf die hydraulischen Bodeneigenschaften und die zugrunde liegende Porencharakteristik untersucht. In-situ Messungen mittels Tensionsinfiltrometer wurden durchgeführt, um Veränderungen der hydraulischen Leitfähigkeit, des wasserleitenden Porenraumes und von invers bestimmten Parametern des van Genuchten/Mualem (VGM) Retentionsmodells zu quantifizieren. Im ersten Teil der Arbeit wurden Infiltrationsmessungen in unterschiedlich stark verdichteten Varianten eines schluffigen Lehmbodens in Lincoln/Neuseeland verwendet, um den Einfluss auf die Porosität und die Wasserleitfähigkeit zu charakterisieren. Es wurde eine hohe Anfälligkeit für die aufgebrachte Verdichtung festgestellt, wobei die gesättigte hydraulische Leitfähigkeit der stärksten Verdichtungsstufe um 81% geringer als im gelockerten Unterboden war. Die gesättigte hydraulische Leitfähigkeit könnte als Proxy für Änderungen der hydraulischen Eigenschaften durch Bodenverdichtung verwendet werden. Zunehmende Verdichtung führte zum Abnehmen der Anzahl hydraulisch wirksamer Makroporen, des flussgewichteten Porenradius' und des α_{VG}-Parameters des VGM Modells. Im zweiten Teil der Arbeit wurde der Einfluss unterschiedlicher Bodenbearbeitung – konventionell (CT), reduziert (RT) und bearbeitungslos (NT) – und dessen zeitliche Dynamik auf einen schluffigen Lehmboden in Raasdorf/Niederösterreich durch wiederholte Messungen mittels Tensionsinfiltrometer erfasst. Die Ergebnisse zeigen, dass die nah-gesättigte hydraulische Leitfähigkeit in der Reihenfolge CT > RT > NT war, wobei größere Unterschiede zwischen den Varianten im Bereich der Mesoporen auftraten. Der VGM Parameter α_{VG} war in der Reihenfolge CT < RT < NT, wobei bei CT und RT eine starke zeitliche Variabilität festgestellt wurde. NT führte zu den größten wasserleitenden Porenradien. Die Ergebnisse geben einen indirekten Hinweis darauf, dass NT zu einer besseren Konnektivität und geringeren Tortuosität der Makropren sowie zu einer höheren zeitlichen Stabilität führt. Die Variabilität im Mesoporenbereich konnte hinreichend durch eine Kombination aus Bodenbearbeitung und Zeit erklärt werden. Im dritten Teil wurden die invers bestimmten hydraulischen Parameter verwendet, um eine Simulation des Bodenwasserhaushaltes für zwei Jahre zu parametrisieren. Simulationen des oberflächennahen Wasserhaushaltes mit konstanten und zeitlich variablen hydraulischen Parametern wurden mit gemessenen Wassergehalten verglichen. Die Verwendung zeitlich variabler Parameter verbesserte die Performance der Simulation signifikant, wobei der durchschnittliche relative Fehler unter 13% lag. Die Untersuchungen zeigen, dass invers bestimmte hydraulische Parameter für Simulationen des Bodenwasserhaushaltes verwendet werden können. Die Simulation des Bodenwasserhaushaltes deutet an, dass RT und NT im Gegensatz zu CT zu einer erhöhten Wasserspeicherung im Oberboden führen. Damit wird die Wassereffizienz besonders unter trockeneren Klimabedingungen gesteigert.

Schlüsselwörter: hydraulische Bodeneigenschaften, hydraulische Leitfähigkeit, Porengrößenverteilung, Tensionsinfiltrometer, Bodenverdichtung, Bodenbearbeitung, zeitliche Variabilität

1. Introduction

The movement and storage of water and solutes in soils is of fundamental importance to answer recent questions in agricultural water management and for the assessment of cultivation practices. As agriculture has to face increasing water shortages and soil degradation in many areas worldwide, there is a need for assessing the impact of different landuse changes and agricultural practices on the hydraulic properties, the pore network, and the water balance of soils. In particular, the effect of different tillage techniques and management-induced subsoil compaction on the soil hydraulic properties are of interest, not only in fundamental research but also for stakeholders and farmers. It has been reported that soil tillage and management affect the hydraulic properties with consequences for the storage and movement of water, nutrients and pollutants, and for plant growth (Strudley et al., 2008). Many studies have demonstrated that soil tillage and compaction alter the soil pore-size distribution and the hydraulic properties (Mubarak et al., 2009; Or et al., 2000; Xu and Mermoud, 2003).

For the assessment of agricultural impacts on the movement and storage of water in the soil, numerical simulations can be a helpful tool. However, the hydraulic properties, i.e. the soil water retention function $\theta(h)$ and the hydraulic conductivity function $K(h)$, have to be properly known. Generally, one main challenge in the assessment of management-induced impacts on these properties is the high natural spatial and temporal variability, particularly in the near-saturated range where soil structure essentially influences water flow characteristics (Daraghmeh et al., 2008; Or et al., 2000). As classical methods to derive the soil hydraulic properties from steel core samples cover only a small soil volume and its variability, adequate measurements techniques should be applied. Moreover, most of the management-induced changes are expected to occur in the structural soil pores. This part of the soil's porosity can be captured best by using field measurements (Angulo-Jaramillo et al., 2000; Hu et al., 2009; Yoon et al., 2007). To determine the hydraulic properties of agricultural soils directly in the field, tension infiltrometry has become a commonly used method (Angulo-Jaramillo et al., 1997; Messing and Jarvis, 1993; Reynolds et al., 1995). This method is the basis of the following thesis. It allows not only to determine the saturated and near-saturated hydraulic conductivity, but also to derive the pore-size distribution that contributes to the water infiltration and conductance. By applying an inverse parameter estimation, tension infiltrometer measurements can be also used to derive the parameters of a soil water retenion and conductivity model, such as the van Genuchten/Mualem model (van Genuchten, 1980).

The results of the present thesis might contribute to a better understanding of management-induced changes of soil hydraulic properties with respect to the temporal dynamics and its implications for the agricultural practice – having the aim to prevent soil degradation and optimize water efficiency.

2. Hypotheses and objectives

This thesis follows the motivation to contribute research results to the scientific and agricultural community with the objective of a good agricultural practice. The following main hypotheses were tested:
- Subsurface compaction alters the soil's capability for water infiltration, with the major impact on fast-draining macropores.
- Increasing soil compaction results in a decrease in the near-saturated hydraulic conductivity.
- Different tillage techniques might result in different pore size distributions and thus soil hydraulic properties.
- Soil hydraulic properties vary with time as a result of both different soil cultivation techniques and environmental controlling factors.
- The impact of soil compaction and tillage on the soil hydraulic properties can be captured using tension infiltrometer measurements.
- Soil water simulations can be improved by accounting for temporal changes of the soil hydraulic properties.

The hypotheses resulted in the following main research objectives:
- Assessing the impact of subsoil compaction on the near-saturated hydraulic conductivity and the water-conducting porosity.
- Assessing the impact of different tillage techniques and their subsequent temporal dynamics on the near-saturated hydraulic conductivity and derived soil hydraulic properties.
- Implementation of time-variable soil hydraulic properties in a soil water simulation.
- Assessing the feasibility of inversely estimated hydraulic parameters in soil water simulations.

3. Structure of the study

This doctoral thesis consists of three independent chapters. In chapter 5, the impact of subsoil compaction on the hydraulic properties and the water-conducting porosity is discussed. Field measurements for this study were conducted on a silt loam soil in Lincoln, New Zealand. In chapter 6, the impact of different tillage practices on the hydraulic properties and the water-conducting porosity is analyzed with respect to its temporal dynamic. Over a period of two consecutive years, repeated infiltration measurements were conducted on a silt loam soil in Raasdorf, Austria. This data was further analyzed and used to parameterize a soil water simulation that accounts for time-variable hydraulic properties (chapter 7).

4. Dissemination

Since this is a cumulative doctoral thesis, the central parts of this study were subject to scientific publications in peer-reviewed journals (listed in the Science Citation Index by Thomson Reuters, NY). Prior submission for publication, the results were presented and discussed on international scientific conferences. Details are listed in Tables 4-1 and 4-2.

Table 4-1. List of publications in journals that are listed in the Science Citation Index (SCI).

Chapter of this study	SCI-Journal	Impactfactor (2009)	Full reference
5	Soil Science Society of America Journal	2.18	Schwen, A., Hernandez-Ramirez, G., Lawrence-Smith, E.J., Sinton, S.M., Carrick, S., Clothier, B.E., Buchan, G.D., and Loiskandl, W. 2011. Soil Sci. Soc. Am. J. 75(3), 822-831.
6	Soil & Tillage Research	2.88	Schwen, A., Bodner, G., Scholl, P., Buchan, G.D., and Loiskandl, W. 2011. Soil & Tillage Research 113(2), 89-98.
7	Agricultural Water Management	2.02	Schwen, A., Bodner, and Loiskandl, W. 2011. Agricultural Water Management 99, 42-50.

Table 4-2. List of presentations on international scientific conferences.

Chapter of this study	Conference	Date & City	Title and type of presentation
5	European Geosciences Union - General Assembly 2010	May 2-7, 2010 Vienna, Austria	Using Tension Infiltrometry to Assess the Effect of Subsoil Compaction on Soil Hydraulic Properties (oral)
6	European Geosciences Union - General Assembly 2010	May 2-7, 2010 Vienna, Austria	Assessing Tillage Effects on Soil Hydraulic Properties via Inverse Parameter Estimation using Tension Infiltrometry (Poster)
7	19th World Congress of Soil Science	Aug 1-6, 2010 Brisbane, Australia	The Effect of Time-Variable Soil Hydraulic Properties in Soil Water Simulations (Poster)

5. Hydraulic Properties and the Water-Conducting Porosity as Affected by Subsurface Compaction using Tension Infiltrometers

Abstract

Changes in soil physical properties due to compaction are a major concern in agricultural production and modeling of soil water movement and plant growth. The objectives of this study were to measure water infiltration under different compaction levels and to characterize the effects of compaction on the soil's porosity and its associated water conducting properties. On a silt loam soil, relative to a control four levels of subsurface compaction were induced: loosening, light, medium and heavy compaction. Infiltration characteristics were measured *in-situ* using tension infiltrometers. Near-saturated conductivities were calculated using Wooding's equation and further analyzed to derive the hydraulically effective macro- and mesoporosities and the flow-weighted mean pore radii. The data were also used for parametrization of the van Genuchten/Mualem model by inverse parameter estimation. A high susceptibility to the applied compaction was found, as the saturated hydraulic conductivity of the heavy compacted soil was 81% less than that of the loosened soil. This soil property may be used as a proxy for compaction-induced changes in hydraulic characteristics. Increasing compaction also decreased the number of hydraulically effective macropores, reduced the flow-weighted mean pore radius, and the α_{VG} parameter of the van Genuchten/Mualem model. Our findings give an indirect evidence for two main effects of the applied compaction. Firstly, the reduced saturated hydraulic conductivity might be due to distortion of structural flow paths, reducing the connectivity and hydraulic effectiveness of many macropores. Secondly, compaction rearranged the pore space, resulting in more water-conducting mesopores.

5.1 Introduction

Globally, compaction is a major cause of soil degradation that is mainly associated with agricultural practices or forest harvesting. Soil compaction alters the soil's structure by crushing aggregates or combining them into larger units and by decreasing the number of coarser pores and reducing water and air permeability (Batey, 2009; Batey and McKenzie, 2006). In humid areas, soil compaction might increase the risk of surface runoff and erosion due to a decreased rainwater infiltration (Lipiec and Hatano, 2003). Soil structural degradation and the formation of tillage pans under cultivated cropping are effects that have been widely reported (Shepherd et al., 2001; Hamilton-Manns et al., 2002; Sparling and Schipper, 2004). Fine-textured soils are especially susceptible to compaction (Hewitt and Shepherd, 1997; Gebhardt et al., 2009). Consequently, crop growth, yield and quality may be impaired by reductions in rooting depth, and reduced water and nutrient uptake (Batey and McKenzie, 2006).

Therefore, changes in the soil physical properties due to compaction are growing in importance in agricultural production and research. Soil compaction, through its effects on permeability and water retention, might also affect important input parameters used in predictive crop growth models (e.g. Allen et al., 1998). For instance, reduced porosity could diminish the water content at field capacity and therefore reduce the plant available water capacity (Batey, 2009). Also, the saturated hydraulic conductivity, K_s could decrease due to a reduction in the number of large and fast-draining pores (macropores). Most of the decrease in water retention and conductance is expected to occur at hydraulic pressure heads close to zero due to a reduction in different groups of macropores (Gebhardt et al., 2009).

Many studies have assessed the impact of soil compaction on hydraulic properties using small soil core samples in the laboratory (e.g. Richard et al., 2001; Startsev and McNabb, 2001; Lipiec and Hatano, 2003; Gebhardt et al., 2009; Reintam et al., 2009). However, Startsev and McNabb (2001) noted that compaction-induced changes in soil water retention and pore-size distribution are still not well documented. The effects of tillage-induced compaction on soil hydraulic properties have also not been examined; existing reports focus only on soil surface compaction induced by wheel traffic. Thus, there is a need for *in-situ* measurements of cultivation pan-induced changes of the hydraulic characteristics.

Angulo-Jaramillo et al. (2000) noted that, especially in structured macroporous soils, field methods to evaluate hydraulic properties are preferable to laboratory methods. To quantify the macroporosity and determine the near-saturated hydraulic properties of soils directly in the field, tension infiltrometers have become a commonly used method (Angulo-Jaramillo et al., 2000; Ramos et al., 2006; Moret and Arrúe, 2007). With this device, water infiltration rates at certain adjustable supply pressure heads are measured. The data can be used to obtain the hydraulic conductivity function using Wooding's analytical equation (Wooding,

1968; Reynolds and Elrick, 1991). Subsequently, estimates of the near-saturated hydraulic conductivities can be further analysed in terms of the amount of hydraulically effective macro- and mesopores (Watson and Luxmoore, 1986; Bodhinayake et al., 2004), as well as to estimate the flow-weighted mean pore radii (Reynolds et al., 1995; Moret and Arrúe, 2007) or to estimate inversely the parameters of a soil-water retention model (Šimůnek et al., 1998). Previous studies have compared the inverse parameter estimation from tension infiltrometer data with retention data obtained from laboratory measurements, and the method has been found to be sound (Angulo-Jaramillo et al., 2000; Schwarz and Evett, 2002; Ramos et al., 2006; Yoon et al., 2007; Daraghmeh et al., 2008).

Our main hypothesis was that subsurface compaction alters the soil's capability for water infiltration, with the major impact being on the infiltration through fast-draining macropores. We expected a decrease in the near-saturated hydraulic conductivity, which could be measured best using the tension infiltrometer technique. Hence, the objective of the study was to assess the impact of subsoil compaction on the near-saturated hydraulic conductivity and the water-conducting porosity. *In-situ* infiltration measurements were obtained on the surface and the variously compacted subsurface of a silt loam soil. Subsequently, the infiltration data was used to estimate the hydraulic conductivity properties, derive changes in the water-conducting pore sizes, and determine the retention characteristics using inverse parameter estimation.

5.2 Materials and Methods

5.2.1 Experimental Site

Field measurements were obtained in an arable field north of Lincoln, Canterbury, New Zealand (43°64′S 172°50′E), with a Templeton silt loam over gravels, classified as a Dystric Ustochrept in the US Soil Taxonomy (Soil Survey Staff, 2010). The particle size analysis resulted in 0.23 kg kg^{-1} sand, 0.52 kg kg^{-1} silt, and 0.25 kg kg^{-1} clay. The soil's organic carbon content, determined using the method of Nelson and Sommers (1982), was 27 g kg^{-1} in a depth of 0.05 m and 24 g kg^{-1} in a depth of 0.30 m.

Treatments were imposed during October 2009. In all treatments, the uppermost 0.15–0.20 m of soil was removed prior to treatment application and replaced afterwards using a digger with a 7.5 m reach. At all stages, care was taken to ensure the digger's caterpillar tracks did not enter the plot measurement area. The subsoil was either left untreated (U) or cultivated with a tyne implement to 0.1 m (loosened, L), or compacted using a heavy roller (10 Mg roller, total length 5.6 m, drum width 2.1 m) with three different intensities of compaction (C1: One pass with heavy roller, 'light compaction'; C2: 8 passes with heavy roller, 'medium compaction'; C3: 8 passes with the roller set to vibration, 'heavy compaction'). This type of compaction was applied to simulate a laterally homogeneous, dense tillage-pan formation at a depth where it typically occurs (Hamilton-Manns et al., 2002; Capowiez et al., 2009). It is acknowledged that this roller method for inducing compaction may not have the same physical effect on the subsoil as regular soil cultivation where compaction of the subsoil would be induced by moldboard plowing or using power harrows.

The five treatments were arranged in a latin square design with five replicates (Figure 5-1). Each plot was 20 m long and 3 m wide. Plots within a replicate were 0.5 m apart and replicates were 17 m apart. The field was sown with barley (*Hordeum vulgare* L.) on November 5th, 2009. To quantify the applied compaction, penetration resistance (*PR*) was measured within 0 to 0.40 m depth using a cone penetrometer (1.28 cm cone diameter, 30° cone angle; Field Scout SC-900, Spectrum Technologies, Inc., Plainfield, IL) on December 7th, 2009.

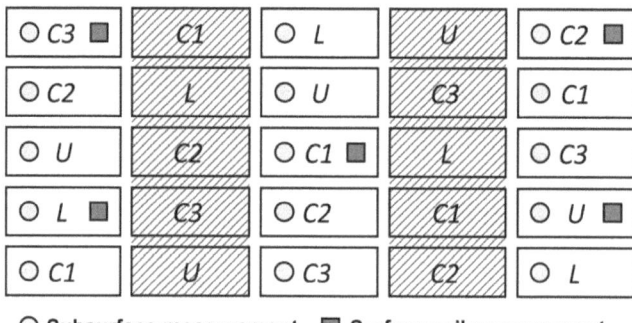

Figure 5-1. Plots of the subsurface compaction trial arranged in a latin square and locations of the infiltration measurements. Within each measured plot, three replicate measurements were conducted. The treatments were: U: untreated subsoil, L: loosened subsoil, C1: light compaction, C2: medium compaction, C3: heavy compaction. The hatched areas were restricted for other samplings within the research project.

5.2.2 Tension Infiltration Measurements

Infiltration was measured during February 2010, at crop maturity. The measurements were performed using three tension infiltrometers (Soil Measurement Systems Inc., Tucson, AZ) of the design described by Ankeny et al. (1988). The infiltrometer disc had a diameter of 0.20 m and was separate from the supply and tension control tubes. Before each measurement, the soil surface was prepared by carefully removing mulch and any aboveground plant material. For the subsurface measurements, the topsoil was excavated to a depth of 0.15–0.20 m, according to the actual depth of the compacted soil layer. The soil surface for infiltration was carefully leveled using a trowel. We placed a nylon mesh on the smoothed soil to protect any macropores and used a 2 mm layer of uniform glass beads between the disc and the soil to ensure an adequate hydraulic contact (Reynolds and Zebchuk, 1996). The supply pressure heads were −12, −9, −6, −3, −1, and 0 cm. The first two pressure heads were maintained for approximately 60 min, the following two for about 40 min, and the last pressure heads were applied for about 10–20 min. Preliminary tests found these durations to be sufficient to achieve steady-state infiltration. The water level in the water supply tube was measured using a pressure transducer (Casey and Derby, 2002). The transducers of three infiltrometers were connected to a CR 10X data logger (Campbell Scientific Ltd., Logan, UT), and readings were recorded at intervals of 30 s.

For the four subsurface compaction treatments and the control, infiltration measurements were made on three of the five field replicates, as the remaining plots were restricted for other samplings within the research project (Figure 5-1). Within each plot, three replicate measures were conducted ($n = 9$ per treatment). Soil surface measurements were not done on a treatment basis; we randomly selected five experimental plots across all assessed treatments and replicates, with three replicate measures in each plot ($n = 15$, Figure 5-1). The large number of replicate measurements was made to account for the high spatial variability of field-saturated water flow parameters (Warrick and Nielsen, 1980).

Before each set of infiltration measurements, disturbed soil samples were taken with steel cores (diameter: 100 mm, height: 50 mm) in the vicinity of the measurement location in order to obtain the initial water content gravimetrically. Immediately after each infiltration measurement, another core sample was collected directly below the infiltration disc at a shallow depth to quantify the final water content and the bulk density gravimetrically.

5.2.3 Wooding's Equation

We used the procedure described by Reynolds and Elrick (1991) and Ankeny et al. (1991) to determine the saturated and near-saturated hydraulic conductivity $K(h)$. The variably saturated water flow equation can be analytically approximated for infiltration from a circular source with a constant pressure head at the soil surface and with the unsaturated hydraulic conductivity described by the exponential model of Gardner (1958):

$$K(h) = K_s \exp(\alpha_{Ga} h) \tag{5.1}$$

where h is the pressure head [L], $K(h)$ is the unsaturated hydraulic conductivity [L T^{-1}] at pressure head h, K_s is the saturated hydraulic conductivity [L T^{-1}], and α_{Ga} is the sorptive number. The analytical approximation derived by Wooding (1968) is given by:

$$q(h) = \left(\pi r_d^2 + \frac{4 r_d}{\alpha_{Ga}} \right) K(h) \tag{5.2}$$

where q is the steady-state infiltration rate [L^3 T^{-1}], and r_d is the radius of the disc [L]. Since the solution of Wooding has two unknown variables, $K(h_0)$ and α_{Ga}, two steady-state fluxes at different tensions are required. Reynolds and Elrick (1991) derived the unsaturated hydraulic conductivity in the middle of an interval between two applied pressure heads, h_i and h_{i+1}, assuming α_{Ga} to be constant over this interval. If it is also assumed that Eq. [5.1] and [5.2] can be applied piecewise, then:

$$\alpha_{Ga(i+1/2)} = \frac{\ln(q_i / q_{i+1})}{h_i - h_{i+1}}; \quad i = 1, \ldots, n-1 \tag{5.3}$$

where n is the number of supply pressure heads applied. Using Eq. [5.2] gives:

$$K_{i+1/2} = \frac{q_i / q_{i+1}}{1 + [4 / \pi r_d \alpha_{Ga(i+1/2)}]}; \quad i = 1, \ldots, n-1 \tag{5.4}$$

K_s can be calculated from Eq. [5.1] using known values of $h_{i+1/2}$, $K_{i+1/2}$ and $\alpha_{g(i+1/2)}$ as follows:

$$K_s = \frac{K_{i+1/2}}{\exp(\alpha_{Ga(i+1/2)} h_{i+1/2})} \tag{5.5}$$

5.2.4 Hydraulically Effective Porosities and Flow-Weighted Mean Pore Radius

The difference between the infiltration rates at two pressure heads, q_m [L T^{-1}], can be used to estimate the hydraulically effective macro- or mesoporosity of a soil (Buczko et al., 2006). As the method of Watson & Luxmoore (1986) is reported to overestimate the porosities (Bodhinayake et al., 2004; Buczko et al., 2006), we used the more recent approach of Bodhinayake et al. (2004). From the pressure head h the equivalent pore radius r [L] can be calculated using the equation of capillarity (Vomocil, 1965):

$$r = \frac{2\sigma \cos \gamma}{\rho g |h|} \qquad [5.6]$$

Here, σ is the surface tension of water [M T^{-2}], ρ the density of water [M L^{-3}], γ the contact angle between the water-air interface and the solid phase, and $g = 9.81$ m s^{-2} is the acceleration due to gravity [L T^{-2}]. We set $\sigma = 0.0713$ N m^{-1} at 15 °C and assumed γ to be 0°. We note that γ was not determined and might differ distinctly from 0°. Woche et al. (2005) analyzed the dependence between the contact angle and the soil's texture and observed small contact angles of 0–20° for silt loam soils. Moreover, γ strongly depends on the water content and probably approached 0° after a sufficient time of infiltration (Buczko et al., 2006). We used the pore classification of Luxmoore (1981), where macropores have a pressure head range $h \geq -3$ cm and mesopores -300 to -3 cm. Applying Eq. [5.6], these pressure heads correspond to pore radii of $r > 0.5$ mm for macropores and 5×10^{-3} mm $< r < 0.5$ mm for mesopores. The hydraulically active porosity θ_m [L^3 L^{-3}] can be calculated from the hydraulic conductivity in the pressure head interval corresponding to the two pore radii a and b (Bodhinayake et al., 2004):

$$\theta_m(a,b) = \frac{2\eta \rho g}{\sigma^2} \int_{h(a)}^{h(b)} \frac{dK(h)}{dh} h^2 dh \qquad [5.7]$$

Here, η is the dynamic viscosity of water [M L^{-1} T^{-1}], taken here as 0.00115 Pa s for 15 °C. Integration of Eq. [5.7] was done numerically using the inversely parametrized van Genuchten/Mualem model (van Genuchten, 1980) (referred to as VGM) for the mesopore range, and model-independent using cubic-splines for the macropore range. Porosities were calculated as sums of macro- and mesopores and also quasi-continuous over a large pressure-head range following the approach of Carey et al. (2007).

The flow-weighted mean pore radius R_0 [L] is an index that represents an effective equivalent mean pore radius that is conducting water at a certain supply pressure head, and has been used to characterize temporal and tillage-induced changes in water-conducting macropores (Philip, 1985; Sauer et al., 1990; Messing and Jarvis, 1993; Reynolds et al., 1995; Moret and Arrúe, 2007). Following Reynolds et al. (1995), R_0 is defined by:

$$R_0 = \frac{\sigma K_0}{\rho g M_0} \qquad [5.8]$$

Here, M_0 [L^2 T^{-1}] is the matric flux potential of a soil, measured over the pore water pressure head range, where macropores are considered to be water-conducting, and can be calculated by:

$$M_0 = \int K(h)\,dh \qquad [5.9]$$

We integrated the $K(h)$ relationship numerically using cubic splines over the pressure head range of the infiltration measurements. The density of R_0 pores, N_0 [number of pores L^{-2}], can be estimated using the Poiseuille's law relationship:

$$N_0 = \frac{8\mu K_0}{\rho g \pi R_0^4} \qquad [5.10]$$

Applying Eq. [5.6], we also calculated the maximum equivalent pore radius $R_{\max}$, that can be water-conducting at a given supply pressure head (Reynolds et al., 1995; Moret and Arrúe, 2007).

5.2.5 Inverse Parameter Estimation and Calculation of Treatment Means

The inverse analysis of tension infiltrometer data requires a numerical solution of the following modified Richards' equation for radially symmetric Darcy flow (Warrick, 1992):

$$\frac{\partial \theta}{\partial t} = \frac{1}{r}\frac{\partial}{\partial r}\left(rK\frac{\partial h}{\partial r}\right) + \frac{\partial}{\partial z}\left(K\frac{\partial h}{\partial z}\right) + \frac{\partial K}{\partial z} \qquad [5.11]$$

Here, θ is the volumetric water content [L^3 L^{-3}], t the time [T], r is the radial coordinate [L] and z is the vertical coordinate [L], being positive upward with $z = 0$ on the soil surface. The initial and boundary conditions were defined as proposed by Šimůnek et al. (1998). To describe the unsaturated soil hydraulic properties, we used the VGM model that was inversely fitted to the infiltration data. The soil water retention $S_e(h)$ and hydraulic conductivity $K(\theta)$ are given by (van Genuchten, 1980):

$$S_e(h) = \frac{\theta(h) - \theta_r}{\theta_s - \theta_r} = \frac{1}{\left(1 + |\alpha_{VG} h|^n\right)^m} \qquad [5.12]$$

$$K(\theta) = K_s S_e^l \left[1 - \left(1 - S_e^{1/m}\right)^m\right]^2 \qquad [5.13]$$

Here, S_e is the effective water content [-], θ_r and θ_s denote the residual and saturated water contents, respectively [L^3 L^{-3}], l is a pore-connectivity parameter [-], and α_{VG} [L^{-1}], n and m (= 1 − 1/n) are empirical parameters.

We formulated the objective function that is minimized during the parameter estimation using the cumulative infiltration data in combination with the $K(h)$ values calculated by Wooding's equation, and the observed water content at the end of the infiltration experiment θ_f, assuming that this water content corresponds to the final supply pressure head. Minimization of the objective function was accomplished using the Levenberg-Marquardt nonlinear minimization method (Marquardt, 1963), as provided by the program HYDRUS 2D/3D (Šimůnek et al., 2006). For the numerical solution of Eq. [5.11], a quasi-three-dimensional (axisymmetric) model geometry was choosen as described by Šimůnek et al. (1998).

Initial values for the fitting parameters were derived from the soil's texture using the pedotransfer function Rosetta (Schaap et al., 2001; input parameters: soil texture and bulk density). To reduce the number of unknown variables, l was set to a constant value of 0.5 (Ramos et al., 2006), and θ_r was fixed to 0.065 m^3 m^{-3}, as predicted from the pedotransfer analysis of the soil's texture (Lazarovitch et al., 2007), for all parameter estimations. As there is no unique solution, when K_s and α_{VG} are estimated simultaneously (Schwarz and Evett, 2002), K_s was set to the value obtained by Wooding's equation (Lazarovitch et al., 2007; Yoon et al., 2007). The remaining parameters θ_s, α_{VG}, and n were inversely estimated. As the water retention and conductivity relationships are highly nonlinear (Vereecken et al., 2007), representative mean parameters for each compaction treatment were derived using the scaling approach. From all replicate measurements per treatment, $\theta(h)$ data derived from the inverse parameter estimation and the $K(h)$ data obtained by Wooding's equation were used. Following the approach of Schwärzel et al. (2010) and Vereecken et al. (2007), a conventional scaling procedure was applied in which scaling factors were estimated by minimizing the residual sum of square differences between the data and the scaled $K(h)$ and $\theta(h)$ reference curves.

5.2.6 Statistical Analysis

A one-way analysis of variance (ANOVA) was used to test for statistical significance of differences among means of the measured or calculated quantities. The Kolmogorov-Smirnov test was applied to determine if replicates of measured quantities within a treatment were normally distributed or lognormally distributed. In ANOVA, log-transformed values were used where necessary. The coefficient of variation (CV) was calculated using the method of moments for normally distributed data and the maximum likelihood method for lognormally distributed data (Warrick and Nielsen, 1980; Parkin et al., 1988).

5.3 Results and Discussion

5.3.1 Soil Compaction and Infiltration Measurements

The applied subsurface compaction resulted in a slight statistically significant (P < 0.05) increase of the bulk density ρ_b and the penetration resistance *PR* between the treatments L, C1, C2, and C3. However, both quantities showed hardly any differences between treatments L and C1.
During all of our infiltration measurements, steady-state infiltration was reached shortly after the supply pressure heads were applied. This suggests a sufficient duration of the first applied pressure head (Schwartz and Evett, 2002). For all treatments, the cumulative infiltration increased markedly at supply pressures of $h \geq -3$ cm, and this is attributed to activation of soil macropores that have a much greater conductivity than the pores in the soil's matrix. Since the used tension infiltrometer disc is a laterally unconfined device, infiltration rates at $h = 0$ cm could not be measured properly due to leakage, and were excluded from the data analysis. Thus, for Wooding's analysis we extrapolated from the prior infiltration data by assuming logarithmic scales for both pressure heads and infiltration rates (Buczko et al., 2006). While tension infiltrometers have become a commonly accepted method for measuring the hydraulic properties of topsoils (e.g. assessment of different tillage methods), few studies have used this device in subsoils (e.g. Schwartz and Evett, 2002; Garg et al., 2009). The moderate variability in infiltration measurements across replicates in our study confirms the value of the tension infiltrometer for determining the characteristics of excavated and compacted subsoils.

5.3.2 Near-Saturated Hydraulic Conductivity

To assess the impact of subsoil compaction on the near-saturated hydraulic conductivity, we analyzed the inferred $K(h)$ values and their variability. Geometric mean values for $K(h)$ are given in Table 5-1. The saturated conductivity K_s ($h = 0$) showed a strong decrease by one order of magnitude due to the applied soil compaction. The geometric mean values decreased by 81.2% between the treatments L and C3, with 2.06×10^{-3} cm s^{-1}, 1.45×10^{-3} cm s^{-1}, 1.38×10^{-3} cm s^{-1}, and 0.39×10^{-3} cm s^{-1} for treatments L, C1, C2, and C3, respectively. The mean value for the untreated soil was 1.09×10^{-3} cm s^{-1}. However, only the difference in the K_s value of the heaviest compaction C3 was statistically significant at $P < 0.05$. With more negative pressure heads, differences between the treatment means of $K(h)$ receded and were not statistically different. These findings are in agreement with other studies that analyzed the impact of compaction on soils with loamy texture (Startsev and McNabb, 2001; Richard et al., 2001). The $K(h)$ values of the surface soil were approximately twice the corresponding value of the untreated subsoil, and agreed very well with results of Schwärzel et al. (2010) and Jarvis and Messing (1995). We found only a weak relationship between the K_s values and the corresponding values of ρ_b (R^2 = 0.49) or PR (R^2 = 0.52) (Table 5-2). Gebhardt et al. (2009) reported the same finding regarding the bulk density for sandy soils. Since measurements of bulk density contain no information regarding pore geometry and continuity, we agree with Sparling and Schipper (2004) that bulk density cannot be used as a proxy, on its own, for the effects of subsoil compaction. As K_s can be measured easily with many field- and laboratory methods, we propose this quantity to be used as a better proxy for soil compaction in terms of the hydraulical impact.

Table 5-1. Unsaturated hydraulic conductivity $K(h)$ as obtained from Wooding's equation of the tension infiltrometer measurements. Geometric mean values per treatment are shown. The coefficient of variation (CV) of the lognormally-distributed values was calculated using the maximum likelihood method (Parkin et al., 1988).

Treatment	h= –12 cm	h= –10.5 cm	h= –7.5 cm	h= –4.5 cm	h= –2 cm	h= –0.5 cm
			cm s^{-1} × 10^{-5}			
Surface soil (S)	1.33 n.s. (25.3) ‡	2.03 n.s. (26.2)	6.00^a † (17.9)	16.33^a (14.4)	58.38^a (24.0)	128.16^a (21.0)
Untreated (U)	0.82 n.s. (44.6)	1.36 n.s. (33.4)	3.20^b (17.1)	7.16^b (34.5)	23.40^{a,b} (58.6)	31.06^b (27.7)
Loosened (L)	1.22 n.s. (17.7)	1.92 n.s. (21.1)	2.00^b (33.9)	7.26^{b,c} (24.8)	27.07^b (39.6)	74.90^{a,b} (46.7)
Light compaction (C1)	0.91 n.s. (21.9)	1.29 n.s. (18.7)	4.19^b (16.5)	5.45^c (19.5)	19.87^b (30.4)	59.25^b (37.9)
Med. compaction (C2)	1.05 n.s. (29.3)	1.83 n.s. (18.0)	3.12^b (15.7)	6.34^c (25.6)	19.00^b (24.7)	71.37^{a,b} (19.9)
Heavy compaction (C3)	0.94 n.s. (13.2)	1.48 n.s. (14.2)	2.40^b (41.9)	3.63^c (30.2)	9.66^b (22.3)	21.90^b (25.7)

† Different letters indicate significant differences among treatments (P < 0.05)
‡ Values in parenthesis denote the coefficient of variation, CV (%)

5.3.3 Hydraulically Effective Porosity

The hydraulically effective macroporosity correlates to the pressure head range that could be measured completely by the tension infiltrometer. Therefore, this quantity could be calculated model-independent using cubic splines to describe the $K(h)$ relationship. As the correlating pressure-head range of mesopores ($h = -300$ to -3 cm) exceeds the lower limit of the tension infiltrometer measurements ($h = -12$ cm), this quantity could not be calculated using cubic splines, but only by using a model describing the $K(h)$ relationship. To assess the models, we compared the macroporosities derived by the use of cubic splines to the $K(h)$ models of Gardner (1958) and the VGM model. We found good agreement between the model-independent values and the values using the VGM model with the parameters derived from the inverse parameter estimation. Thus, we used the VGM $K(h)$ relationship to calculate the hydraulically effective mesoporosities.

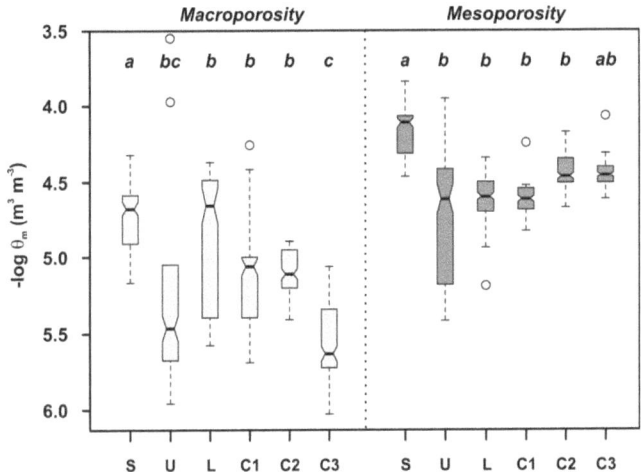

Figure 5-2. Hydraulically effective macroporosity (left) and mesoporosity (right) according to the method of Bodhinayake et al. (2004) using the pore classification of Luxmoore (1981) with $r > 0.5$ mm for macropores and 5×10^{-3} mm $< r < 0.5$ mm for mesopores. S: surface soil, U: untreated subsoil, L: loosened subsoil, C1: light compaction, C2: medium compaction, C3: heavy compaction. Boxes denote the 25th and 75th percentile positions, the line inside the box shows the median value. The whiskers denote the 10th and 90th percentiles, the circles denote outliers, and different letters indicate significant differences among treatments ($P < 0.05$).

The sums of hydraulically effective macropores showed a decrease with increasing subsoil compaction in the geometric mean values with 1.30×10^{-5}, 1.10×10^{-5}, 1.28×10^{-5}, and 0.49×10^{-5} m^3 m^{-3} for the treatments L, C1, C2, and C3, respectively (Figure 5-2).

However, only the decrease in the heaviest compaction treatment C3 was statistically significant (P < 0.05) and resulted in a reduction of 82 % compared to the L treatment. These results are in the same range as those reported by other studies (Table 5 in Buczko et al., 2004). By contrast, there was a slight, statistically not significant increase in hydraulically effective mesopores with increasing soil compaction with 2.26×10^{-5}, 2.56×10^{-5}, 3.63×10^{-5}, and 3.78×10^{-5} $m^3\ m^{-3}$ for the treatments L, C1, C2, and C3, respectively. Compared to the porosities of the subsoil, the surface soil contained a greater proportion of hydraulically effective mesopores.

The change in porosities can be better understood when the quantity of a particular water-conducting pore-size is calculated continuously over the whole pressure-head range (Figure 5-3a). Compared to the untreated control (U), the loosening of the subsoil (L) resulted in an decreased quantity of water-conducting mesopores, with a peak water-conducting porosity in the macropore range at $r_m = 1.33$ mm. With increasing soil compaction the water-conducting porosity shifted towards smaller pore radii. Although the maximum water-conducting porosity for the treatment C1 also peaked at $r_m = 1.33$ mm, the trend towards smaller pore radii becomes evident when the number of water-conducting pores N_m is analyzed (Figure 5-3b). With increasing compaction, the number of smaller pores increased and the maximum N_m value of water-conducting pores is then found in the mesopore range.

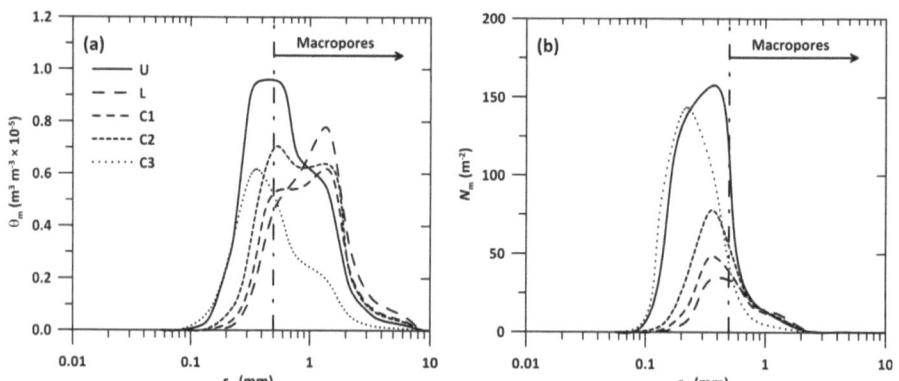

Figure 5-3. Water-conducting porosity θ_m (a) and number of hydraulically effective pores N_m (b) for different pore radii r_m. The vertical dashed line separates the macropores from the mesopores. U: untreated subsoil, L: loosened subsoil, C1: light compaction, C2: medium compaction, C3: heavy compaction. Surface soil data are not shown.

The decrease of hydraulically effective macropores with increasing compaction was supported by an statistically significant (P < 0.05) increase in the mean bulk density (Table 5-2). However, as macroporosity is only a small part of the total porosity, the decrease in hydraulically effective macropores might not be explained by a loss in total macroporosity. To investigate further, we calculated the flow-weighted mean pore radius R_0 as an alternative method for characterizing the pore-size distribution in terms of its hydraulic behavior. This approach showed that the mean pore radius conducting water at a certain supply pressure head during the tension infiltration experiment decreased with increasing soil compaction (Figure 5-4a). The maximum equivalent pore radius, R_{max}, and R_0 were approximately equal at $h \leq -9$ cm. For the more saturated conditions, there were increasing differences between R_{max} and R_0. As stated by Reynolds et al. (1995), R_0, compared to storage-based R_{max}, better reflects the effects of pore restrictions, such as entrapped air bubbles or small unwetted zones. As the differences between R_{max} and R_0 increased with increasing soil compaction, we can infer that soil compaction resulted in a reduction of the pore connectivity, leaving a certain fraction of the macropores disconnected from the water-conducting pores. The largest R_0 value that becomes water conducting at $h = 0$ decreased with increasing soil compaction with 0.85, 0.68, 0.62, and 0.30 mm for the treatments L, C1, C2, and C3, respectively (Figure 5-4b). This observation supports the trend towards smaller water-conducting pore radii (Figure 5-3). The decrease in R_0 was largest for the C3 treatment. Compared to the other treatments, the N_0 vs. R_0 relationship is highly skewed (Figure 5-4b), with the maximum N_0 in the vicinity of $R_0 = 0.16$ mm. This shows that the application of a vibrating roller caused greater soil compaction and distortion of the pore space, thereby leaving hardly any undamaged macropores. As there was only a small increase in ρ_b and PR between treatments C2 and C3 (Table 5-2), only the infiltration measurements revealed the strong reduction of fast-draining macropores in the C3 treatment.

The analysis of the water-conducting porosity and the flow-weighted mean pore radius give evidence that increasing compaction might have resulted in two main effects. Firstly, the distortion of structural flow paths such as earthworm burrows and root channels might have reduced the connectivity and hydraulic effectiveness of many macropores through the so-called bottleneck effect. Secondly, as a fraction of these pores remains disconnected and hydraulically inactive, the compaction had probably induced a rearrangement of the pore space, resulting in an increase in hydraulically effective mesopores. These hypotheses agree with those in other studies (Richard et al., 2001; Gebhardt et al., 2009). However, we note that changes in the soil's pore structure have not been measured directly but have been derived from the infiltration measurements indirectly.

The Kolmogorov-Smirnov test revealed that replications of $K(h)$ and θ_m were lognormally distributed, showing the inherently high natural spatial variability of pore characteristics. These findings are in agreement with those in other studies (Parkin et al., 1988; Mesquita et al., 2002; Buczko et al., 2006). Thus, the variabilities of these quantities are discussed in

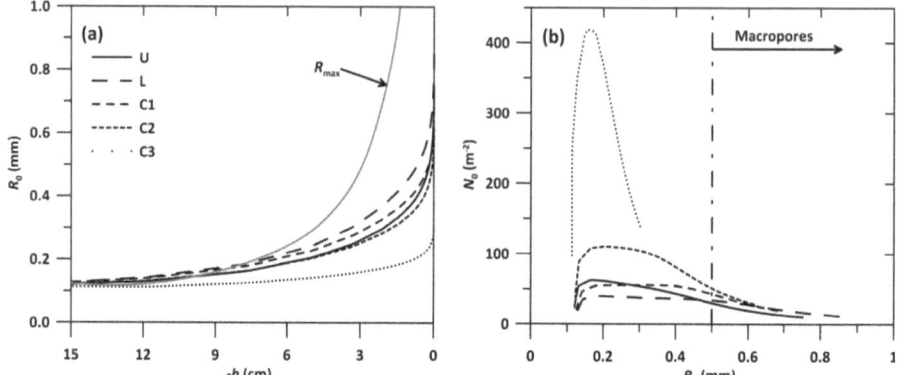

Figure 5-4. a: Flow-weighted mean pore radius R_0 versus pressure head h in the range of the tension infiltrometer measurements. R_{max} denotes the maximum equivalent pore radius. b: Number of flow-weighted mean pores N_0 versus flow-weighted mean pore radius R_0. The vertical dashed line separates the macropores from the mesopores. U: untreated subsoil, L: loosened subsoil, C1: light compaction, C2: medium compaction, C3: heavy compaction. Surface soil data are not shown.

terms of the CV of the log-transformed values. Variability in θ_m and K_s within the treatments was largest in the untreated subsoil and generally larger in the macropore range (Table 5-1, Figure 5-2). These findings agree with results of other authors who reported decreasing CV with increasing negative pressure heads (Watson and Luxmoore, 1986; Carey et al., 2007). However, our calculated CVs, with a maximum of 117 % for K_s of the U treatment and most smaller than 30 %, were quite low compared to published ranges for mineral soils (Warrick and Nielsen, 1980; Watson and Luxmoore, 1986; Reynolds et al., 2000). The higher CV near saturation might be attributed to a high natural spatial variability in the presence or absence of water-conducting macropores and preferential flow paths. The variability was smaller where the subsoil was artificially altered, either loosened or compacted. The lower variability in the loosened treatment may be attributed to a homogenisation of the soil structure by tillage. This hypothesis is supported by an increase in macropore abundance (Figure 5-2), indicating that mechanical loosening created a finer structure with a more spatially linked macropore network. The applied compaction also decreased the variability due to distortion of macropores and the re-packing of aggregates.

These changes in the spatial variability also suggest that the natural distribution of macropores in the untreated soil is on a coarser scale than the dimensions of the infiltrometer disc can cater for. In this sense, the subsoil treatments that were applied may have reduced the characteristic length of variation of the macropore network (Jarvis and Messing, 1995), resulting in a size that better matches the size of the disc.

Table 5-2. Physical properties within the compacted soil layer and results of the infiltration measurements for different steps of subsoil compaction. The results of Wooding's equation are the saturated hydraulic conductivity K_s and the sorptive number α_{Ga}. The inverse parameter estimation gave the water content at saturation θ_s and the shape parameters of the VGM model α_{VG} and n.

Treatment	Surface soil (S)	Untreated (U)	Loosened (L)	Light compaction (C1)	Medium compaction (C2)	Heavy compaction (C3)
ρ_b g cm^{-3}	1.01^a † (4.0) ‡	1.25^b (5.6)	1.24^b (3.2)	1.24^b (4.0)	1.32^c (10.6)	1.34^c (4.5)
PR MPa	n.d.	1.80^{a,b}	1.61^a	1.69^a	2.29^b	2.35^b
Wooding's analysis						
K_s ×10^{-5} cm s^{-1}	261.0^a (27.0)	109.5^{a,b,c} (117.1)	205.7^b (59.8)	144.7^b (60.2)	138.4^b (23.2)	38.6^c (39.4)
α_{Ga} m^{-1}	66.7^{a,b} (15.1)	51.4^{a,b} (97.1)	84.8^a (32.9)	74.2^a (52.8)	66.1^{a,b} (44.5)	41.3^b (66.1)
Inverse parameter estimation						
θ_s m^3 m^{-3}	0.349^a (15.2)	0.320^{a,b} (9.7)	0.339^a (6.8)	0.297^b (12.8)	0.292^b (15.1)	0.285^b (12.3)
α_{VG} m^{-1}	18.5^{a,b} (6.4)	19.3^{a,b} (31.2)	25.4^a (15.8)	20.8^{a,b} (21.1)	17.3^{b,c} (7.9)	8.6^c (12.5)
n -	1.646^a (16.6)	1.426^b (11.2)	1.49^{a,b} (9.5)	1.501^{a,b} (12.5)	1.464^{a,b} (4.8)	1.478^{a,b} (24.4)

† Different letters indicate significant differences among treatments (P < 0.05)
‡ Values in parenthesis denote the coefficient of variation, CV (%)

The infiltration experiments were carried out during February 2010, and they reveal the state of compaction that remained after the barley vegetative growth period. The impact can be expected to be even larger shortly after the compaction was imposed. Especially, biological activities such as root growth and earthworm burrows could have altered the initial levels of compaction. Recently, Capowiez et al. (2009) studied the distribution of earthworm activity in compacted soils. They observed that earthworms play a significant role in the regeneration of compacted soils. The species *Lumbricus terrestris* is known to cross the plough pan in vertical burrows that can remain as hydraulically effective macropores. This biological induced regeneration of the applied compaction and its temporal dynamic could not be covered by our measurements. We recommend that future investigations focus on temporal changes in soil hydraulic properties following soil compaction, with a series of infiltration measurements following the application of the compaction and throughout the growing season.

5.3.4 Inverse Parameter Estimation

Our inverse parameter estimation allowed the determination of the VGM model parameters for all infiltration experiments. Representative mean values were derived for each treatment (Table 5-2). We note that the infiltration measurements were done starting at relatively high initial water contents of 0.15–0.22 $m^3 m^{-3}$. This resulted in a quite small range between the initial and final water content of the infiltration measurements, which might not allow a reliable estimation of the VGM parameters. However, the VGM model is a closed-form set of equations, where the retention curve and the unsaturated conductivity curve are described by the same parameters (Eq. [5.12] and [5.13]; van Genuchten, 1980). Since the $K(h)$ values obtained by Wooding's equation were included in the objective function during the inverse parameter estimation, a good description of the conductivity relationship and thus also of the VGM parameters was enabled.

The applied subsoil compaction resulted in a reduction in θ_s. However, only the L treatment was statistically significant (P < 0.05). As the total porosity of the analyzed soil was in the order of 0.48 $m^3 m^{-3}$ (data not shown), our values for θ_s were considerably lower. As reported by Šimůnek et al. (1998), the water content at saturation θ_s is governed by the final water content θ_f and is typically 5-10 % lower when measured under field-wetted conditions than obtained by laboratory methods, where a soil sample gets fully saturated in a water bath, resulting in less entrapped air in soil pores.

To evaluate the performance of the inverse parameter estimation, the shape parameters of the VGM model were compared with values derived from the pedotransfer function Rosetta (Schaap et al., 2001; input parameters: soil texture and bulk density). The observed n value from the surface measurement (1.64) was in the range as expected for sandy loam soils, whereas the values of the subsurface measurements (1.43-1.50) were in the range of a silty loam soil, (1.43 reported by Jarvis and Messing, 1995). The estimated α_{VG} values (9-25 m^{-1}) were slightly larger than expected for silt loam soils (7 m^{-1} using Rosetta), and agreed better with values for silty clay soils (21-24 m^{-1} using Rosetta). However, as the inversely fitted α_{VG} values were in the same order of magnitude as expected from the soil's texture, the applied inverse parameter estimation approach gave reliable results. These findings agree with other studies that compared inversely estimated retention data with water release experiments (Schwarz and Evett, 2002; Ramos et al., 2006; Yoon et al., 2007). However, in structured soils values of α_{VG} and n can vary between the macropore- and mesopore-range. Jarvis and Messing (1995) reported an inverse relationship for these parameters between the macropore-range and the soil's matrix. Thus, we acknowledge that extrapolation of hydraulical parameters measured in the near-saturated range should be made carefully and not too far beyond the measured range.

Our results indicate, that α_{VG} was affected by the subsoil compaction, as the representative mean values decreased statistically significant ($P < 0.05$) from 25.4 m^{-1} to 8.6 m^{-1} between the treatments L and C3. As we found no statistically significant differences between the subsurface treatments, we conclude that n was hardly affected by the applied compaction (Table 5-2). However, we acknowledge that the inverse parameter estimation of n shows only weak sensitivity to the near-saturated infiltration measurements (Deurer, 2000).

5.4 Conclusion

Our study reveals the applicability of tension infiltrometers to determine the hydraulic properties and quantify changes in the water-conducting macro- and mesoporosity in differentially compacted subsoils. The *in-situ* application of different levels of subsurface compaction affected the hydraulic properties, reducing K_s and the hydraulically effective macroporosity θ_m as well as the flow-weighted mean pore radius R_0 and the α_{VG} parameter of the VGM model. As a result, we suggest the easy-measurable K_s value to be used as a proxy for the hydraulical impact of soil compaction.

We conclude that the analyzed silt loam soil is highly susceptible to subsurface compaction in terms of a loss in fast-draining macropores and the continuity of the pore network. The results suggest that parts of the water-conducting macropores seem to be transformed into conducting mesopores in response to the applied compaction. These findings may be explained by distortion and reduction of the connectivity of the macropore network due to the applied compaction. The reduced water infiltration may increase water logging, surface runoff and soil erosion during heavy precipitation events. The applied soil compaction may also negatively affect biological activities in the soil and the gas exchange between the subsoil and the atmosphere. As a consequence for crop growth and soil water modeling purposes, we suggest accounting for a reduced K_s in compacted subsurface layers.

5.5 References

Allen R.G., L.S. Pereira, D. Raes, and M. Smith. 1998. Crop evapotranspiration – Guidelines for computing crop water requirements. FAO Irrigation and drainage paper 56. (Food and Agriculture Organization of the United Nations. Rome).

Angulo-Jaramillo, R., J.P. Vandervaere, S. Roulier, J.L. Thony, J.P. Gaudet, and M. Vauclin. 2000. Field measurement of soil surface hydraulic properties by disc and ring infiltrometers. A review and recent developments. Soil & Tillage Research 55:1–29.

Ankeny, M.D., T.C. Kaspar, and R. Horton. 1988. Design for an automated tension infiltrometer. Soil Sci. Soc. Am. J. 52:893–896.

Ankeny, M.D., M. Ahmed, T.C. Kaspar, and R. Horton. 1991. Simple field method for determining unsaturated hydraulic conductivity. Soil Sci. Soc. Am. J. 55:467–470.

Batey, T. and D.C. McKenzie. 2006. Soil compaction: identification directly in the field. Soil Use and Management 22:123–131.

Batey, T. 2009. Soil compaction and soil management – a review. Soil Use and Management 25:335–345.

Buczko, U., O. Bens, and R.F. Hüttl. 2006. Tillage effects on hydraulic properties and macroporosity in silty and sandy soils. Soil Sci. Soc. Am. J. 70:1998–2007.

Bodhinayake, W., B.C. Si, and C. Xiao. 2004. New method for determining water-conducting macro- and mesoporosity from tension infiltrometer. Soil Sci. Soc. Am. J. 68:760–769.

Capowiez, Y., S. Cadoux, P. Bouchand, J. Roger-Estrade, G. Richard, and H. Boizard. 2009. Experimental evidence for the role of earthworms in compacted soil regeneration based on field observations and results from a semi-field experiment. Soil Biology & Biochemistry 41:711–717.

Carey, S.K., W.L. Quinton, and N.T. Goeller. 2007. Field and laboratory estimates of pore size properties and hydraulic characteristics for subarctic organic soils. Hydrol. Process. 21:2560–2571.

Casey, F.X.M. and N.E. Derby. 2002. Improved design for an automated tension infiltrometer. Soil Sci. Soc. Am. J. 66:64–67.

Daraghmeh, O.A., J.R. Jensen, and C.T. Petersen. 2008. Near-saturated hydraulic properties in the surface layer of a sandy loam soil under conventional and reduced tillage. Soil Sci. Soc. Am. J. 72:1728–1737.

Deurer, M. 2000. The dynamics of water and solute flow in the unsaturated zone of a coniferous forest: Measurement and numerical simulation. Herrenhaeuser Contributions to Soil Science, Volume 2, University of Hannover, Germany.

Gardner, W.R. 1958. Some steady-state solutions of the unsaturated moisture flow equation with application to evaporation from a water table. Soil Sci. 85:228–232.

Garg, K.K., B.S. Das, M. Safeeq, and P.B.S. Bhadoria. 2009. Measurement and modeling of soil water regime in a lowland paddy field showing preferential transport. Agricultural Water Management 96:1705–1714.

Gebhardt, S., H. Fleige, and R. Horn. 2009. Effect of compaction on pore functions in soils in a Saalean moraine landscape in North Germany. J. Plant Nutr. Soil Sci. 172:688–695.

Hamilton-Manns, M., C.W. Ross, D.J. Horne, and C.J. Baker. 2002. Subsoil loosening does little to enhance the transition to no-tillage on a structurally degraded soil. Soil & Tillage Research 68:109–119.

Hewitt, A.E. and T.G. Shepherd. 1997. Structural vulnerability of New Zealand soils. Australian J. Soil Res. 35:461–474.

Jarvis, N.J. and I. Messing. 1995. Near-saturated hydraulic conductivity in soils of contrasting texture measured by tension infiltrometers. Soil Sci. Soc. Am. J. 59:27–34.

Lazarovitch, N., A. Ben-Gal, J. Šimůnek, and U. Shani. 2007. Uniqueness of Soil Hydraulic Parameters Determined by a Combined Wooding Inverse Approach. Soil Sci. Soc. Am. J. 71(3):860–865.

Lipiec J. and R. Hatano. 2003. Quantification of compaction effects on soil physical properties and crop growth. Geoderma 116:107–136.

Luxmoore R.J. 1981. Micro- meso- and macroporosity of soil. Soil Sci. Soc. Am. J. 45:671–672.

Marquardt, D.W. 1963. An algorithm for least-quares estimation of nonlinear parameters. SIAM J. Appl. Math. 11:431–441.

Mesquita, M., S.O. Moraes, and J.E. Corrente. 2002. More adequate probability distributions to represent the saturated soil hydraulic conductivity. Scientia Agricola 59(4):789–793.

Messing, I. and N.J. Jarvis. 1993. Temporal variation in the hydraulic conductivity of a tilled clay soil as measured by tension infiltrometers. J. Soil Sci. 44:11–24.

Moret, D. and J.L. Arrue. 2007. Characterizing soil water-conducting macro- and mesoporosity as influenced by tillage using tension infiltrometry. Soil Sci. Soc. Am. J. 71(2):500–506.

Nelson, D.W. and L.E. Sommers. 1982. Total carbon, organic carbon, and organic matter. p. 539–579. *In* A.L. Page et al. (ed.) Methods of soil analysis. Part 2. Chemical and microbiological properties. Agron. Monogr. 9. ASA and SSSA, Madison, WI.

Parkin, T.B., J.J. Meisinger, S.T. Chester, J.L. Starr, and J.A. Robinson. 1988. Evaluation of statistical estimation methods for lognormally distributed variables. Soil Sci. Soc. Am. J. 52:323–329.

Philip, J.R. 1985. The quasilinear analysis, the scattering analog, and other aspects of infiltration and seepage. p. 1–27. *In:* Infiltration development and application, Y.-S. Fok (ed.) Water Resources Research Center, Honolulu,

Ramos, T.B., M.C. Gonçalves, J.C. Martins, M.Th. van Genuchten, and F.P. Pires. 2006. Estimation of soil hydraulic properties from numerical inversion of tension disk infiltrometer data. Vadose Zone J. 5:684–696.

Reynolds, W.D. and D.E. Elrick. 1991. Determination of hydraulic conductivity using a tension infiltrometer. Soil Sci. Soc. Am. J. 55:633–639.

Reynolds, W.D., E.G. Gregorich, and W.E. Curnoe. 1995. Characterisation of water transmission properties in tilled and untilled soils using tension infiltrometers. Soil & Tillage Research 33:117–131.

Reynolds, W.D. and W.D. Zebchuk. 1996. Use of contact material in tension infiltrometer measurements. Soil Technology 9:141–159.

Reynolds, W.D., B.T. Bowman, R.R. Brunke, C.F. Drury, and C.S. Tan. 2000. Comparison of tension infiltrometer, pressure infiltrometer, and soil core estimates of saturated hydraulic conductivity. Soil Sci. Soc. Am. J. 64:478–484.

Reintam, E., K. Trükmann, J. Kuht, E. Nugis, L. Edesi, A. Astover, M. Noormets, K. Kauer, K. Krebstein, and K. Rannik. 2009. Soil compaction effects on soil bulk density and penetration resistance and growth of spring barley (*Hordeum vulgare* L.). Acta Agriculturae Scandinavica, Section B – Plant Soil Science, 59(3):265–272.

Richard, G., I. Cousin, J.F. Sillon, A. Bruand, and J. Guérif. 2001. Effect of compaction on the porosity of a silty soil: influence on unsaturated hydraulic properties. Eur. J. Soil Sci. 52:49–58.

Sauer, T.J., B.E. Clothier, and T.C. Daniel. 1990. Surface measurements of the hydraulic properties of a tilled and untilled soil. Soil Tillage Res. 15:359–369.

Schaap, M.G., F.J. Leij, and M.Th. van Genuchten. 2001. ROSETTA: a computer program for estimating soil hydraulic parameters with hierarchical pedotransfer functions. J. Hydrology 251(3–4):163–176.

Schwärzel, K., Carrick, S., Wahren, A., Feger, K.-H., Bodner, G., and Buchan, G.D., 2011. Soil hydraulic properties of recently tilled soil under cropping rotation compared with 2-years-pasture: Measurement and modelling the soil structure dynamics. Vadose Zone Journal 10(1), 354-366.

Schwartz, R.C. and S.R. Evett. 2002. Estimating hydraulic properties of a fine-textured soil using a disc infiltrometer. Soil Sci. Soc. Am. J. 66:1409–1423.

Shepherd, T.G., S. Saggar, R.H. Newman, C.W. Ross, and J.L. Dando. 2001. Tillage-induced changes to soil structure and organic carbon fractions in New Zealand soils. Aust. J. Soil Res. 39:465–489.

Šimůnek, J., R. Angulo-Jaramillo, M.G. Schaap, J.P. Vandervaere, and M.Th. van Genuchten. 1998. Using an inverse method to estimate the hydraulic properties of crusted soils from tension-disc infiltrometer data. Geoderma 86:61–81.

Šimůnek, J., N.J. Jarvis, M.Th. van Genuchten, and A. Gärdenäs. 2003. Review and comparison of models for describing non-equilibrium and preferential flow and transport in the vadose zone. J. Hydrology 272:14–35.

Šimůnek, J., M.Th. van Genuchten, and M. Šejna. 2006. The HYDRUS Software Package for Simulating the Two- and Three-Dimensional Movement of Water, Heat, and Multiple Solutes in Variably-Saturated Media. Technical Manual. Version 1.0. PC Progress, Prague, Czech Republic.

Soil Survey Staff. 2010. Keys to soil taxonomy, 11th ed. USDA-Natural Resources Conservation Service, Washington, DC.

Sparling, G. and L. Schipper. 2004. Soil quality monitoring in New Zealand: trends and issues arising from a broad-scale survey. Agriculture Ecosystems & Environment 104:545–552.

Startsev, A.D. and D.H. McNabb. 2001. Skidder traffic effects on water retention, pore-size distribution, and van Genuchten parameters of boreal forest soils. Soil Sci. Soc. Am. J. 65:224–231.

van Genuchten, M.Th. 1980. A closedform equation for predicting the hydraulic conductivity of unsaturated soils. Soil Sci. Soc. Am. J. 44:892–898.

Vereecken, H., R. Kasteel, J. Vanderborght, and T. Harter. 2007. Upscaling hydraulic properties and soil water flow processes in heterogeneous soils: a review. Vadose Zone J. 6:1–28.

Vomocil, J.A. 1965. Porosity. p. 299–314. *In* C.A. Black (ed.) Methods of soil analysis. Part 1. Agron. Monogr. 9, ASA, Madison, WI.

Warrick, A.W. 1992. Model for disc infiltrometers. Water Resour. Res. 28:1319–1327.

Warrick, A.W. and D.R. Nielsen. 1980. Spatial variability of soil physical properties in the field. *In*: D. Hillel (ed.) Applications of soil physics. Academic Press, Toronto, Canada.

Watson, K.W. and R.J. Luxmoore. 1986. Estimating macroporosity in a forest watershed by use of a tension infiltrometer. Soil Sci. Soc. Am. J. 50:578–582.

Woche, S.K., M.O. Goebel, M.B. Kirkham, R. Horton, R.R. van der Ploeg, and J. Bachmann. 2005. Contact angle of soils as affected by depth, texture, and land management. European J. Soil Sci. 56:239–251.

Wooding, R.A. 1968. Steady infiltration from a shallow circular pond. Water Resour. Res. 4:1259–1273.

Yoon, Y., J.G. Kim, and S. Hyun. 2007. Estimating soil water retention in a selected range of soil pores using tension disc infiltrometer data. Soil & Tillage Res. 97:107–116.

6. Temporal dynamics of soil hydraulic properties and the water-conducting porosity under different tillage

Abstract

Soil hydraulic properties are subject to temporal changes as a response to both tillage and natural impact factors. As the temporal and spatial variability might exceed cultivation-induced differences, there is a need to better differentiate between those influence factors. Thus, the objective of the present study was to assess the impact of different tillage techniques – conventional (CT), reduced (RT), and no-tillage (NT) – on the soil hydraulic properties and their temporal dynamics. On a silt loam soil, tension infiltrometer measurements were obtained frequently over two consecutive years. The data was analyzed in terms of the near-saturated hydraulic conductivity, inversely estimated parameters of the van Genuchten/Mualem (VGM) model, and the water-conducting porosity. Our results show that the near-saturated hydraulic conductivity was in the order CT > RT > NT, with larger treatment-induced differences where water flow is dominated by mesopores. The VGM model parameter α_{VG} was in the order CT < RT < NT, with high temporal variations under CT and RT, whereas the parameter n was hardly affected. NT resulted in the greatest water-conducting pore radii, whereas no distinct differences were observed between CT and RT. The results give indirect evidence that NT leads to greater connectivity and smaller tortuosity of macropores, possibly due to a better established soil structure and biological activity. NT also resulted in a better temporal stability of both the pore network and the hydraulic properties, but showed the highest spatial variability of macropores. We suggest that the hydraulically effective pores decreased after tillage in response to rainfall during winter, and gradually increased in spring and summer induced by biological activity, root development and wetting / drying cycles. A multivariate ANOVA revealed that variations in mesopore-related quantities could be explained sufficiently by an interaction of tillage and time. By contrast, due to high spatial variability, macropore-related quantities could not be explained by those influence factors. The study reveals the importance of the temporal dynamics for both hydraulic properties and the water-conducting porosity.

6.1 Introduction

Soil cultivation practices affect soil hydraulic properties and processes dynamically in space and time with consequences for the storage and movement of water, nutrients and pollutants, and for plant growth (Strudley et al., 2008). For given climatic conditions and a particular soil-plant system, both the tillage practice and irrigation system can alter the soil structure (Messing and Jarvis, 1993). Thus, understanding the temporal and management-induced changes that soil hydraulic properties undergo is important for sound land management and for modeling nutrient or contaminant transport (Angulo-Jaramillo et al., 1997; Hu et al., 2009; Yoon et al., 2007).

The impact of different cultivation techniques on soil hydraulic properties has been frequently studied in recent decades (Ndiaye et al., 2007; Sauer et al., 1990; Strudley et al., 2008). However the literature clearly shows that the impact of different tillage pratices on soil hydraulic properties is not consistent across locations, soils, and experiment designs (Moret and Arrue, 2007a; Strudley et al., 2008). Moreover, most publications document averaged comparisons between different tillage practices and do not take into account the spatial and temporal dynamics (Angulo-Jaramillo et al., 1997; Strudley et al., 2008). Only few published studies address both the temporal and management-induced changes in soil hydraulic properties (Alletto and Coquet, 2009; Angulo-Jaramillo et al., 1997; Bormann and Klaassen, 2008; Cameira et al., 2003; Daraghmeh et al., 2008; Messing and Jarvis, 1993; Moret and Arrue, 2007a). However it was recently published that temporal variability of soil hydraulic properties can exceed differences induced by crops, tillage, or land use (Alletto and Coquet, 2009; Bodner et al., 2008; Hu et al., 2009; Zhou et al., 2008). Although the importance of this variability was demonstrated, it still has not been well studied, and changes in the underlying water-conducting porosity have not yet been referred to in published studies (Hu et al., 2009; Mubarak et al., 2009).

In general, water flow in structured soils is mainly conducted by macropores and larger mesopores even though they constitute only a small fraction of the total porosity (Moret and Arrue, 2007a; Mubarak et al., 2009; Reynolds et al., 1995; Sauer et al., 1990). As the main impact of different tillage techniques on soil hydraulic properties is expected to occur in the structural pores, due to changes in different groups of macro- and mesopores, field methods are preferable to laboratory methods (Angulo-Jaramillo et al., 2000; Hu et al., 2009; Yoon et al., 2007). To determine the near-saturated hydraulic properties of agricultural soils directly in the field, tension infiltrometry has become a commonly used method (Angulo-Jaramillo et al., 1997; Messing and Jarvis, 1993; Reynolds et al., 1995). The measurement of infiltration rates at pre-set supply pressure heads can be used to obtain the hydraulic conductivity function using Wooding's analytical equation (Reynolds and Elrick, 1991; Wooding, 1968). Subsequently, estimates of the hydraulic conductivity can be further analysed to obtain the amount of hydraulically effective macro- and mesopores (Bodhinayake et al., 2004; Watson and Luxmoore, 1986), to estimate the flow-weighted mean pore radius (Moret and Arrue, 2007b; Reynolds et al., 1995), or to estimate inversely the parameters of a soil-water retention model (Šimůnek and van Genuchten, 1996; Šimůnek and van Genuchten, 1997).

Our main hypothesis of this research was that soil hydraulic properties and water-conducting porosities vary with time as a result of both different soil cultivation techniques and environmental controlling factors. The objective of this study was to assess the impact of different tillage techniques and their subsequent temporal dynamics on the near-saturated hydraulic conductivity and derived soil hydraulic properties. The dynamics is then analyzed in terms of underlying changes in the water-conducting porosity.

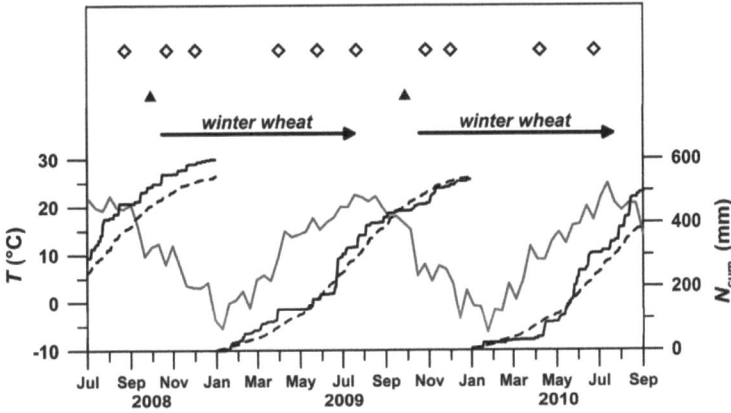

Figure 6-1. Climatic conditions, soil cultivation and times of measurements: 10-d average air temperature T (grey line); cumulative precipitation N_{cum} (black continous line); 10-year mean cumulative precipitation (dotted line). Also shown are times of infiltration measurements (diamonds), soil tillage (triangles), and the crop growing periods.

6.2 Materials and Methods

6.2.1 Experimental Site

Measurements were obtained on an arable field near Raasdorf, Lower Austria (48°14'N 16°35'E). The site has a semi-arid (pannonic) climate with annual average precipitation of 546 mm, and temperature of 9.8 °C, and a mean relative humidity of 75 %. Weather data were recorded at the site since 1998 by an automated weather station (Adcon Telemetry GmbH, Austria) (Figure 6-1).
A field trial was established in 1997 to assess the following soil cultivation techniques: 1) Conventional tillage (CT) with moldboard ploughing and seedbed preparation using a harrow prior crop seeding; 2) Reduced tillage (RT) with a chisel plough to 10 cm for seedbed preparation; and 3) No-tillage (NT) using direct seeding technique. The three treatments were arranged in a randomized complete block design with three replicate plots, each of length 40 m and width 24 m.
The soil can be classified as Chernozem in the WRB (IUSS, 2007) or as Typic Vermudoll in the US Soil Taxonomy (Soil Survey Staff, 2010). Particle size analysis gave 0.27 kg kg^{-1} sand, 0.54 kg kg^{-1} silt, and 0.19 kg kg^{-1} clay, classified as silt loam according to the FAO classification (FAO, 1990). The organic carbon content, determined using the method of Nelson and Sommers (1982), was 24 g kg^{-1} in the topsoil. In both years, the site was cropped with winter wheat (*Triticum aestivum*) in mid October (Figure 6-1).

6.2.2 Tension Infiltrometer Measurements

Infiltration measurements were made ten times between August 2008 and June 2010 (Figure 6-1), using three tension infiltrometers (Soil Measurement Systems Inc., Tucson, AZ) of the design described by Ankeny et al. (1988). The infiltrometer disc (diameter 0.20 m) was separate from the supply and tension control tubes. Before each measurement, the soil surface was prepared by carefully removing mulch and any above-ground plant material. We placed a nylon mesh on the smoothed soil to protect any macropores and used a 2 mm layer of quartz sand (diameter: 0.08-0.2 mm) between disc and soil to ensure good hydraulic contact. The supply pressure heads were −10, −4, −1, and 0 cm: the first two were maintained for approximately 50-60 min, and the last two for about 10-20 min. Preliminary tests found these durations to be sufficient to achieve steady-state infiltration. The water level in the water supply tube was observed visually at intervals of 20 s during the first 5 min after application of a supply pressure, and then in increasing intervals of 2-10 min.

For each treatment, three replicate infiltration measurements were carried out. All experiments were conducted in the non-trafficked interrow area of a plot.

Before each infiltration measurement, soil samples were taken with steel cores near the measurement location to obtain the initial water content θ_i [L³ L⁻³]. Immediately after each measurement, another core sample was collected directly below the infiltration disc to quantify the final water content θ_f [L³ L⁻³], bulk density ρ_b [M L⁻³], and total porosity φ [L³ L⁻³] (Table 6-1).

6.2.3 Data Analysis Procedure

All infiltration experiments were analyzed by the following procedure. First, the near-saturated hydraulic conductivity vs. pressure head relationship $K(h)$ [L T⁻¹] was derived from each single experiment using the analytical approach of Wooding (1968). Afterwards, these $K(h)$ data, together with the cumulative infiltration data and the θ_f, were used to inversely parametrize the van Genuchten/Mualem soil water retention model (van Genuchten, 1980; referred to as VGM model). Representative means of parameters for each treatment at each time of measurement were derived by scaling. The data was further analyzed in terms of the water conducting porosity and flow-weighted mean pore radius.

6.2.4 Near-Saturated Hydraulic Conductivity

We used the procedure described by Reynolds and Elrick (1991) and Ankeny et al. (1991) to determine $K(h)$. The variably saturated water flow equation can be analytically approximated for infiltration from a circular source with a constant pressure head at the soil surface, and with $K(h)$ described by Gardner's exponential model (1958):

$$K(h) = K_s \exp(\alpha_{Ga} h) \qquad [6.1]$$

Here h is the pressure head [L], K_s is the saturated hydraulic conductivity [L T⁻¹], and α_{Ga} is the sorptive number. The analytical approximation derived by Wooding (1968) is:

$$q(h) = \left(\pi r_d^2 + \frac{4 r_d}{\alpha_{Ga}} \right) K(h) \qquad [6.2]$$

where q is the steady-state infiltration rate [$L^3 T^{-1}$], and r_d is the radius of the disc [L]. Since Wooding's solution has two unknown variables, $K(h_0)$ and α_{Ga}, two steady-state fluxes at different tensions are required. Reynolds and Elrick (1991) among others derived $K(h)$ in the middle of an interval between two applied pressure heads h_i and h_{i+1}, assuming α_{Ga} to be constant over this interval. Assuming that Eq. [6.1] and [6.2] can be applied piecewise, then

$$\alpha_{Ga(i+1/2)} = \frac{\ln(q_i / q_{i+1})}{h_i - h_{i+1}}; \quad i = 1,...,n-1$$

[6.3]

where n is the number of applied supply tensions. Using Eq. [6.2] gives:

$$K_{i+1/2} = \frac{q_i / q_{i+1}}{1 + [4/\pi r_d \alpha_{Ga(i+1/2)}]}; \quad i = 1,...,n-1$$

[6.4]

K_s can be calculated from Eq. [6.1] using known values of $h_{i+1/2}$, $K_{i+1/2}$ and $\alpha_{Ga(i+1/2)}$ as follows:

$$K_s = \frac{K_{i+1/2}}{\exp(\alpha_{Ga(i+12)} h_{i+1/2})}$$

[6.5]

6.2.5 Inverse Estimation of Soil Hydraulic Parameters

The inverse analysis of tension infiltrometer data requires numerical solution of the following modified Richards' equation for radially symmetric Darcian flow (Warrick, 1992):

$$\frac{\partial \theta}{\partial t} = \frac{1}{r} \frac{\partial}{\partial r}\left(rK \frac{\partial h}{\partial r}\right) + \frac{\partial}{\partial z}\left(K \frac{\partial h}{\partial z}\right) + \frac{\partial K}{\partial z}$$

[6.6]

Here θ is the volumetric water content [$L^3 L^{-3}$], t the time [T], r is the radial coordinate [L] and z is the vertical coordinate [L], being positive upward with $z = 0$ on the soil surface. The initial and boundary conditions were defined as proposed by Šimůnek et al. (1998). To describe the unsaturated soil hydraulic properties, we used the VGM model. The soil water retention $S_e(h)$ and hydraulic conductivity $K(\theta)$ functions are given by:

$$S_e(h) = \frac{\theta(h) - \theta_r}{\theta_s - \theta_r} = \frac{1}{\left(1 + |\alpha_{VG} h|^n\right)^m}$$

[6.7]

$$K(\theta) = K_s S_e^l \left[1 - \left(1 - S_e^{1/m}\right)^m\right]^2$$

[6.8]

Here S_e is the effective water content [-], θ_r and θ_s denote the residual and saturated water contents, respectively [$L^3 L^{-3}$], l is a pore-connectivity parameter [-], and α_{VG} [L^{-1}], n and m (= 1 – 1/n) are empirical parameters.

We formulated the objective function (OF) that is minimized during parameter estimation by combining the cumulative infiltration data, the $K(h)$ values calculated by Wooding's analysis, and θ_f. OF minimization was accomplished using the Levenberg-Marquardt nonlinear minimization method (Marquardt, 1963), as provided by the program HYDRUS 2D/3D (Šimůnek et al., 2006). For numerical solution of Eq. [6.6], a quasi-3D (axisymmetric) model geometry was chosen, as described by Šimůnek et al. (1998).

Initial values for the parameters were derived from the soil's texture using the Rosetta pedotransfer package Rosetta (Schaap et al., 2001; input parameters: soil texture and ρ_b). To reduce the amount of unknown variables, for all parameter estimations l was set constant at 0.5 (Ramos et al., 2006), and θ_r was fixed at 0.065 m^3 m^{-3}, as predicted by Rosetta (Lazarovitch et al., 2007). K_s was set to the value obtained by Wooding's analysis (Lazarovitch et al., 2007; Yoon et al., 2007). The remaining parameters θ_s, α_{VG}, and n were inversely estimated.

As the $\theta(h)$ and $K(h)$ relationships are highly nonlinear (Vereecken et al., 2007), representative mean parameters for each cultivation treatment were derived using the scaling approach. For each replicate measurement, data derived from the inverse parameter estimation and the data obtained by Wooding's equation were used. Following the approach of Schwärzel et al. (2010) and Vereecken et al. (2007), a conventional scaling procedure was applied in which scaling factors were estimated by minimizing the residual sum of square differences between the data and the scaled $K(h)$ and $\theta(h)$ reference curves.

6.2.6 Description of Water-Conducting Porosities

The difference between the infiltration rates at two pressure heads, q_m [L T^{-1}], can be used to estimate the hydraulically effective porosity (Bodhinayake et al., 2004; Buczko et al., 2006; Watson and Luxmoore, 1986). As the method of Watson & Luxmoore (1986) is reported to overestimate the porosities (Bodhinayake et al., 2004; Buczko et al., 2006), we used the more recent approach of Bodhinayake et al. (2004). For a given h the capillary equation gives the equivalent pore radius r [L] (Vomocil, 1965):

$$r = \frac{2\sigma \cos \gamma}{\rho g |h|} \qquad [6.9]$$

Here, σ is the surface tension of water [M T^{-2}], ρ the density of water [M L^{-3}], γ the contact angle between the water-air interface and the solid phase, and $g = 9.81$ m s^{-2} [L T^{-2}]. We set $\sigma = 0.0713$ N m^{-1} at 15 °C and assumed γ to be 0° (Buczko et al., 2006; Schwen et al., 2011). We used the pore classification of Luxmoore (1981), where macropores have a pressure head range $h \geq -3$ cm and mesopores -300 to -3 cm. Applying Eq. [6.9], these correspond to pore radii of $r > 0.5$ mm for macropores and 5×10^{-3} mm $< r < 0.5$ mm for mesopores. The hydraulically active porosity $\varepsilon(a,b)$ [L^3 L^{-3}] can be calculated from the hydraulic conductivity in the pressure head interval corresponding to the two pore radii a and b (Bodhinayake et al., 2004):

$$\varepsilon(a,b) = \frac{2\eta \rho g}{\sigma^2} \int_{h(a)}^{h(b)} \frac{dK(h)}{dh} h^2 dh \qquad [6.10]$$

Here, η is the dynamic viscosity of water [M L^{-1} T^{-1}], taken here as 0.00115 Pa s for 15 °C. Eq. [6.10] was integrated numerically using the inversely parameterized VGM model. Porosities were calculated as sum for macro- and mesopores (ε_{macro} and ε_{meso}).

The flow-weighted mean pore radius R_0 [L] is an index that represents an effective equivalent mean pore radius that is conducting water at a certain supply pressure head, and has been used to characterize temporal and tillage-induced changes in water-conducting macropores (Messing and Jarvis, 1993; Reynolds et al., 1995; Sauer et al., 1990). Following Reynolds et al. (1995), it is defined by:

$$R_0 = \frac{\sigma K_0}{\rho g M_0} \qquad [6.11]$$

Here, M_0 [L^2 T^{-1}] is the matric flux potential of a soil, measured over the pore water pressure head range, where pores are considered to be water-conducting, and can be calculated by:

$$M_0 = \int K(h) dh \qquad [6.12]$$

The density of R_0 pores, N_0 [number of pores L^{-2}], can be estimated using Poiseuille's law:

$$N_0 = \frac{8\mu K_0}{\rho g \pi R_0^4} \tag{6.13}$$

Applying Eq. [6.9], we also calculated the maximum equivalent pore radius R_{max}, that can be water-conducting at a given h (Moret and Arrue, 2007b; Reynolds et al., 1995).

6.2.7 Statistics

A one-way analysis of variance (ANOVA) was used to test for statistical significance of differences among means of all quantities. The Kolmogorov-Smirnov test was applied to determine if replicates of measured quantities within a treatment were normally or lognormally distributed. In ANOVA, log-transformed values were used where necessary. The coefficient of variation (CV) was calculated using the method of moments for normally distributed data and the maximum likelihood method for lognormally distributed data (Parkin et al., 1988; Warrick and Nielsen, 1980).

To differentiate between temporal and management-induced impacts on the measured quantities, a multivariate ANOVA within the General Linear Model procedure was performed. Homogeneity of variance was assumed.

Table 6-1. Bulk density ρ b, total porosity φ, and volumetric water content prior the infiltration measurement θ_i at the different sampling dates. The treatments were CT (conventional tillage), RT (reduced tillage), and NT (no-tillage). Different letters indicate significant differences among treatments (p < 0.05), values in parenthesis denote the coefficient of variation, CV (%).

	Tillage	Aug. 2008	Oct. 2008	Dec. 2008	Apr. 2009	May 2009	Jul. 2009	Oct. 2009	Dec. 2009	Apr. 2010	Jun. 2010	Mean
ρ b g cm^{-3}	CT	1.25^a† (5.9)‡	1.37 n.s. (11.7)	1.36^a (6.5)	1.39^a (4.3)	1.43 n.s. (2.4)	1.39^a (6.0)	1.30^a (2.8)	1.30^a (4.6)	1.32^{ab} (1.6)	1.30 n.s. (5.8)	1.34^a (5.2)
	RT	1.21^a (8.5)	1.39 n.s. (10.5)	1.3^a (10.6)	1.42^{ab} (5.6)	1.4 n.s. (4.5)	1.34^a (5.9)	1.29^a (4.9)	1.25^a (8.2)	1.28^a (2.3)	1.27 n.s. (1.7)	1.32^a (6.3)
	NT	1.4^b (1.3)	1.45 n.s. (2.2)	1.46^b (5.0)	1.5^b (3.1)	1.46 n.s. (3.5)	1.51^b (3.6)	1.42^b (2.7)	1.44^b (3.1)	1.39^b (5.7)	1.33 n.s. (7.3)	1.44^b (3.7)
φ m^3 m^{-3}	CT	0.53^a (1.4)	0.50^a (8.4)	0.49^{ab} (4.5)	0.48^a (5.5)	0.46 n.s. (2.7)	0.48^a (6.4)	0.51^a (2.8)	0.51^a (5.5)	0.50^{ab} (1.9)	0.51 n.s. (5.2)	0.50^a (5.7)
	RT	0.54^a (6.9)	0.47^{ab} (8.5)	0.51^a (3.0)	0.47^{ab} (3.2)	0.47 n.s. (4.2)	0.49^a (3.3)	0.51^a (2.9)	0.53^a (6.4)	0.52^a (2.2)	0.52 n.s. (4.2)	0.50^a (6.4)
	NT	0.47^b (3.0)	0.45^b (2.2)	0.45^b (1.0)	0.43^b (3.9)	0.45 n.s. (3.8)	0.43^b (3.6)	0.46^b (1.6)	0.46^b (3.6)	0.47^b (7.6)	0.50 n.s. (0.8)	0.46^b (5.2)
θ m^3 m^{-3}	CT	0.18^a (1.4)	0.28^a (8.4)	0.32 n.s. (4.5)	0.31 n.s. (5.5)	0.13^a (2.7)	0.3^a (6.4)	0.17^a (2.8)	0.31^a (5.5)	0.38 n.s. (1.9)	0.23 n.s. (5.2)	0.26 n.s. (4.4)
	RT	0.19^a (6.9)	0.32^{ab} (8.5)	0.33 n.s. (3.0)	0.33 n.s. (3.2)	0.14^a (4.2)	0.32^{ab} (3.3)	0.25^b (2.9)	0.43^b (6.4)	0.41 n.s. (2.2)	0.26 n.s. (4.2)	0.3 n.s. (4.5)
	NT	0.27^b (3.0)	0.33^b (2.2)	0.37 n.s. (1.0)	0.32 n.s. (3.9)	0.21^b (3.8)	0.36^b (3.6)	0.27^b (1.6)	0.22^c (3.6)	0.37 n.s. (7.6)	0.27 n.s. (0.8)	0.3 n.s. (3.1)

6.3 Results and Discussion

6.3.1 Soil Physical Properties

Two consecutive years were evaluated showing slight differences in weather conditions (Figure 6-1). The annual precipitation between August 2008 and July 2009 was 547 mm and slightly higher between August 2009 and July 2010 (559 mm). Remarkably, there was hardly any precipitation during April and May 2009. In response to weather variations, θ_i showed significant differences between the sample times (Figure 6-1, Table 6-1). However, only rarely significant ($p < 0.05$) differences were measured between the treatments, with the greatest moisture contents in the NT treatment.

The measured ρ_b values were significantly ($p < 0.05$) higher under NT for most measured dates, with no significant differences between RT and CT (Table 6-1). The overall mean values were 1.34, 1.32, and 1.44 g cm^{-3} for treatments CT, RT, and NT, respectively. Correspondingly, φ was very similar for CT and RT, but significantly ($p < 0.05$) smaller for NT, with overall means of 0.50, 0.50, and 0.46 m^3 m^{-3}, respectively. The greater ρ_b and smaller φ under NT is likely due to natural compaction, as soil in this treatment has not been artificially loosened by tillage for 12 years (Moret and Arrue, 2007a; Soracco et al., 2010; Strudley et al., 2008). The temporal variation of these quantities was small, as the CVs of the overall mean values for ρ_b and φ were in the same range as the CVs at the different sampling times (Table 6-1).

6.3.2 Infiltration Measurements and Near-Saturated Hydraulic Conductivity

Infiltration was measured ten times between August 2008 and June 2010 (dates shown in Figure 6-2). Since the used tension infiltrometer disc is a laterally unconfined device, infiltration rates at $h = 0$ cm could not be measured properly due to leakage, and were excluded from the data analysis. Thus, for the calculation of K_s ($h = 0$) we extrapolated from the prior infiltration data by assuming logarithmic scales for both h and q (Buczko et al., 2006).

To assess the spatial and temporal variability and the impact of soil cultivation, we analyzed the inferred $K(h)$ values (Figure 2a). The Kolmogorov-Smirnov test revealed that replications of $K(h)$ were lognormally distributed, showing the inherently high natural spatial variability (Buczko et al., 2006; Mesquita et al., 2002; Parkin et al., 1988; Schwen et al., 2011). Therefore, the variability is discussed in terms of the CV of log-transformed values. For most sampling dates $K(h)$ was was in the order CT > RT > NT (Figure 2a), which agrees with findings of Moret and Arrue (2007a). The strongest treatment-induced differences occurred in October 2008, whereas no significant differences between the treatments over the whole range of h were observed in June 2010. However, differences of $K(h)$ between treatments were not the same over the range of applied h. Close to field-saturated conditions ($h = -0.5$ and 0 cm), differences in the conductivities were small and hardly significant ($p < 0.05$). With more negative h, greater differences between the treatments occurred. At $h = -10$ cm, the measured $K(h)$ values were greatest for CT, smaller for RT, and significantly smaller for NT during all measurements ($p < 0.05$). This finding agrees with Hu et al. (2009), who reported larger treatment-induced differences at more negative h.

For most treatments and measurement times, the spatial variability, expressed by the CV of replicate measurements, was smallest at $h = -10$ cm and increased towards saturation (Figure 6-2b). This agrees with other studies that reported larger variability in the soil's structural part than in the textural part (Carey et al., 2007; Schaap and Leij, 2000; Schwen et al., 2011; Watson and Luxmoore, 1986). For most measurements, the CVs were smaller than 100 % and thus quite low compared to published ranges for mineral soils (Reynolds et al., 2000; Warrick and Nielsen, 1980; Watson and Luxmoore, 1986). There was no systematic difference between the CVs of the treatments. However, the greatest CVs were observed under NT close to saturation ($h = -0.5$ and 0 cm; October 2008, May 2009, December 2009, and April 2010 in Figure 2b). The high CVs might be explained by the presence or absence of large macropores (e.g. earthworm burrows). Following findings of Jarvis and Messing (1995), their natural distribution is on a coarser scale than the dimensions of the infiltrometer disc. The characteristic length of variation was smaller in the cultivated treatments RT and CT, resulting in a size that better matches the size of the disc.

6.3.3 Soil Hydraulic Properties

Inverse parametrization of the VGM model from near-saturated tension infiltrometer measurements has frequently been used (Ramos et al., 2006; Schwartz and Evett, 2002; Šimůnek et al., 1998). However, in structured soils α_{VG} and n can vary depending on the part of the retention curve where they are determined (structural or textural part). For instance, Jarvis and Messing (1995) reported an inverse relationship for these parameters between the macropore range and the soil's matrix porosity range. Thus, extrapolation of hydraulical parameters measured in the near-saturated range should be made carefully and not too far beyond the measured range (Schwen et al., 2011). Also, the inverse estimation of n shows only weak sensitivity to the variation in near-saturated infiltration (Deurer, 2000). To evaluate the performance of the inverse parameter estimation the resulting parameters were compared with values from the Rosetta pedotransfer package (Schaap et al., 2001) using the soil texture and ρ_b as input parameters.

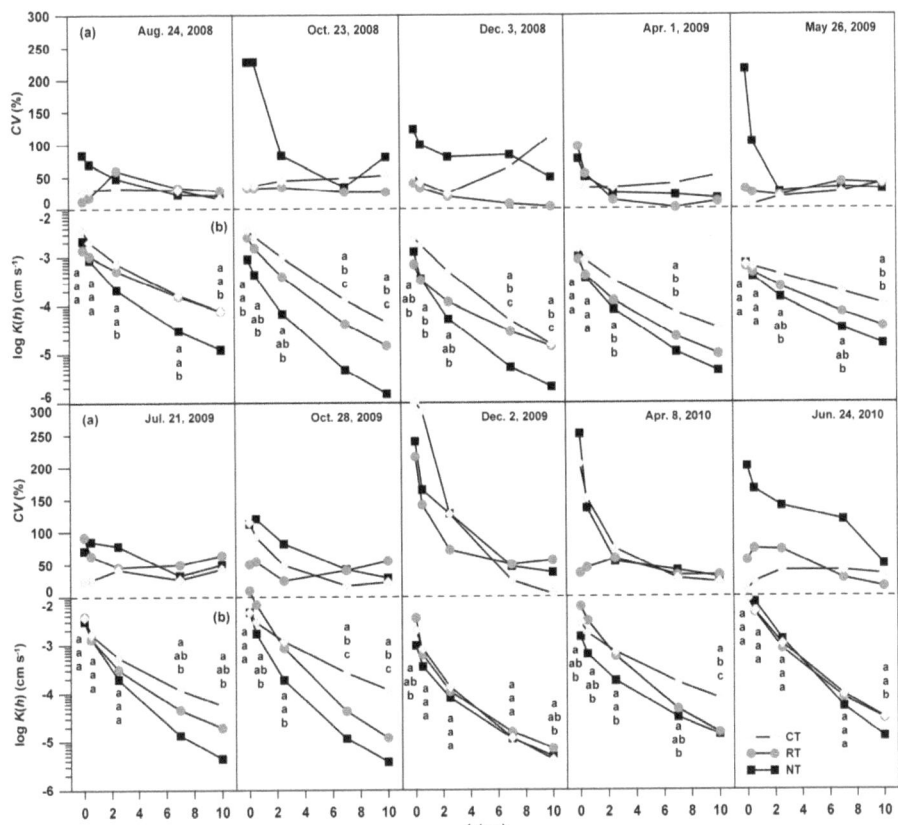

Figure 6-2. (a) Near-saturated hydraulic conductivity $K(h)$ as a function of the supply pressure head h. Representative mean values in the h range of the infiltration measurements for the treatments CT (conventional tillage), RT (reduced tillage), and NT (no-tillage) at each time of measurement are shown. Different letters indicate significant differences among treatments ($p < 0.05$). The lower (b) part of the graphs denotes the coefficient of variation (CV) among replicate measurements within treatments.

θ_s (Table 6-2) agreed with predictions (0.41 m^3 m^{-3}), but was considerably smaller than the corresponding φ value (Table 6-1). It is reported that θ_s is typically 5-10 % lower when measured under field-wetted conditions, due to air entrapment during infiltration (Šimůnek et al., 1998). n ranged between 1.27 and 1.96 with most values around 1.50 and agreed with predictions (1.63). α_{VG} ranged between 6.7 and 42.1 m^{-1} with most values larger than expected for silt loam soils (7 m^{-1}). However, this quantity was highly skewed and lognormally distributed. Since the estimated values were of the same order of magnitude, the inverse parameter estimation gave reliable results. These findings agree with other studies that published inversely estimated hydraulic properties (Ramos et al., 2006; Schwartz and Evett, 2002; Schwen et al., 2011; Yoon et al., 2007).

Differences in θ_s between treatments were hardly significant ($p < 0.05$). For a given date, the CV of the measurements was of the same order for all treatments, indicating no difference in the spatial variability. However, as the CVs of the overall mean values were considerably larger than at the single measurement dates, a temporal variability is indicated for all treatments. θ_s was hardly correlated with the corresponding φ values (regression slope: 0.16, $R^2 = 0.07$), but was correlated well with θ_i (Table 6-1; regression slope: 1.08, $R^2 = 0.45$). This emphasizes the importance of the initial soil moisture content for temporal variation of θ_s.

$K(h)$ may also be influenced by θ_i. Zhou et al. (2008) reported a negative $K(h)$ vs. θ_i relationship for soils under different land use. Thus, we analyzed this correlation for $h = -10$ and 0 cm (Figure 6-3). As no correlation was found for K_s ($h = 0$) for all treatments ($R^2 \leq 0.02$), we suggest that the saturated hydraulic conductivity – representing infiltration mainly through macropores – was not affected by θ_i and its temporal changes. By contrast, K_{10} showed a quite strong correlation for CT ($R^2 = 0.30$) and RT ($R^2 = 0.31$), whereas only weak correlation was found for NT ($R^2 = 0.09$). Thus, infiltration at $h = -10$ cm – where mainly mesopores are water-conducting – was sensitive to changes in θ_i for CT and RT, whereas it was temporally stable for NT.

Table 6-2. Results of the inverse estimation of the van Genuchten/Mualem soil water retention model. Representative mean values at the different sampling dates for the field-saturated water content θ_s and the shape parameters α_{VG} and n are listed. Different letters indicate significant differences among treatments ($p < 0.05$ for θ_s; $p < 0.1$ for α_{VG} and n), values in parenthesis denote the coefficient of variation, CV (%).

	Tillage	Aug. 2008	Oct. 2008	Dec. 2008	Apr. 2009	May 2009	Jul. 2009	Oct. 2009	Dec. 2009	Apr. 2010	Jun. 2010	Mean
θ_s $m^3\,m^{-3}$	CT	0.37 n.s. (1.3) ‡	0.49^a † (2.5)	0.47^a (1.6)	0.41 n.s. (5.2)	0.35^a (5.2)	0.38 n.s. (6.0)	0.43^a (5.4)	0.40^a (2.0)	0.41^a (6.0)	0.38 n.s. (4.4)	0.41 n.s. (10.8)
	RT	0.35 n.s. (8.8)	0.43^b (5.3)	0.42^b (2.5)	0.40 n.s. (4.3)	0.29^b (19.6)	0.39 n.s. (0.9)	0.44^a (11.8)	0.48^b (3.2)	0.47^b (1.8)	0.37 n.s. (5.9)	0.40 n.s. (14.9)
	NT	0.38 n.s. (3.7)	0.41^b (8.9)	0.44^{ab} (2.3)	0.39 n.s. (4.3)	0.29^b (6.3)	0.39 n.s. (3.9)	0.38^b (9.7)	0.34^c (18.5)	0.45^{ab} (9.6)	0.39 n.s. (5.3)	0.39 n.s. (13.0)
α cm^{-1}	CT	0.130 n.s. (7.6)	0.175 n.s. (7.5)	0.217 n.s. (5.4)	0.130 n.s. (9.6)	0.067 n.s. (17.1)	0.139 n.s. (7.3)	0.127 n.s. (8.9)	0.312^a (5.4)	0.120^a (3.6)	0.260 n.s. (6.0)	0.273^a (35.8)
	RT	0.095 n.s. (3.1)	0.224 n.s. (4.0)	0.124 n.s. (4.4)	0.167 n.s. (6.5)	0.094 n.s. (34.1)	0.219 n.s. (7.0)	0.380^{ab} (6.1)	0.285^b (6.3)	0.291^b (2.4)	0.294 n.s. (3.4)	0.329^{ab} (35.7)
	NT	0.216 n.s. (6.8)	0.330 n.s. (4.7)	0.344 n.s. (1.1)	0.236 n.s. (14.8)	0.127 n.s. (13.7)	0.347 n.s. (4.5)	0.421^b (11.5)	0.356^{ab} (54.6)	0.175^{ab} (8.8)	0.325 n.s. (8.6)	0.421^b (27.2)
n -	CT	1.496^a (1.3)	1.809 n.s. (2.5)	1.813^a (1.6)	1.519 n.s. (5.1)	1.858^a (5.1)	1.413^a (5.9)	1.582^a (5.3)	1.549^a (2.0)	1.531 n.s. (6.0)	1.661 n.s. (4.4)	1.812 n.s. (10.6)
	RT	1.651^b (8.7)	1.675 n.s. (5.3)	1.417^b (2.5)	1.464 n.s. (4.3)	1.642^{ab} (18.9)	1.401^a (0.9)	1.748^b (11.6)	1.269^b (3.2)	1.642 n.s. (1.8)	1.509 n.s. (5.9)	1.858 n.s. (16.2)
	NT	1.462^a (3.6)	1.657 n.s. (8.8)	1.418^b (2.3)	1.441 n.s. (4.3)	1.484^b (6.4)	1.621^b (3.9)	1.584^a (9.6)	1.858^a (19.8)	1.434 n.s. (9.7)	1.962 n.s. (5.3)	1.748 n.s. (13.9)

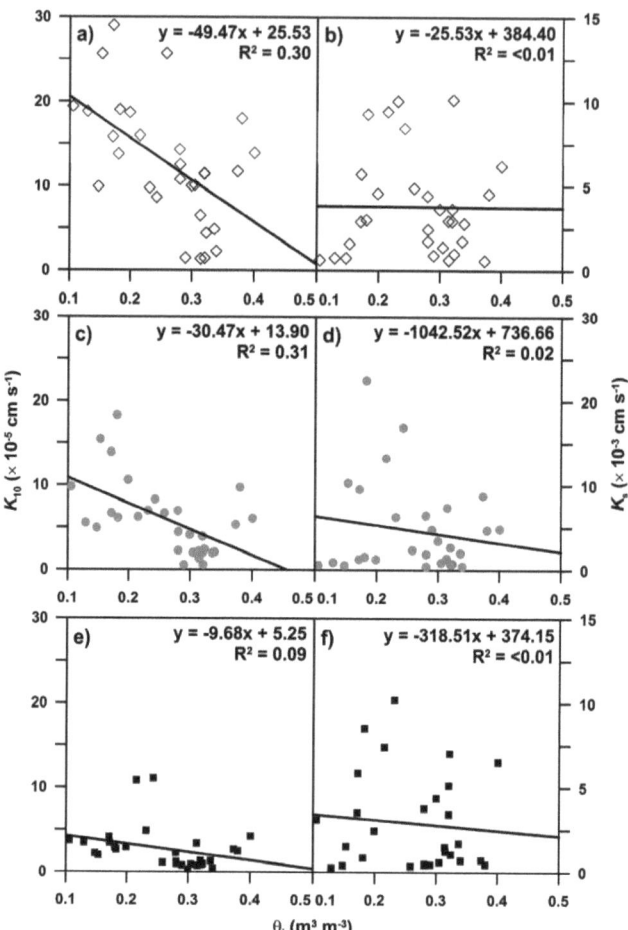

Figure 6-3. Correlation between hydraulic conductivity at $h = -10$ cm (K_{10}) and at saturation (K_s) versus initial volumetric moisture content θ_i for the treatments CT (a and b), RT (c and d), and NT (e and f). A more negative regression slope indicates a higher sensitivity of the hydraulic conductivity to temporal changes in soil moisture (Zhou et al., 2008).

It is acknowledged that any impact of θ_i on the determination of α_{VG} and n could be minimized, as the inverse parameter estimation accounted for θ_i and θ_f. The n values hardly showed significant ($p < 0.1$) differences between treatments and only small temporal dynamic. By contrast, α_{VG} showed significant ($p < 0.1$) differences between treatments with overall means for CT, RT, and NT of 27.3, 32.9, and 42.1 m^{-1}, respectively. This order was found at most measurement dates. However, α_{VG} also showed a considerable temporal dynamic, as indicated by the high CVs of the overall treatment means (Table 6-2). This dynamic was smallest for NT, indicating the relatively greater temporal stability of the hydraulic properties under this cultivation technique. To understand the temporal and management-induced dynamic of α_{VG}, changes in the water-conducting porosity need to be taken into account.

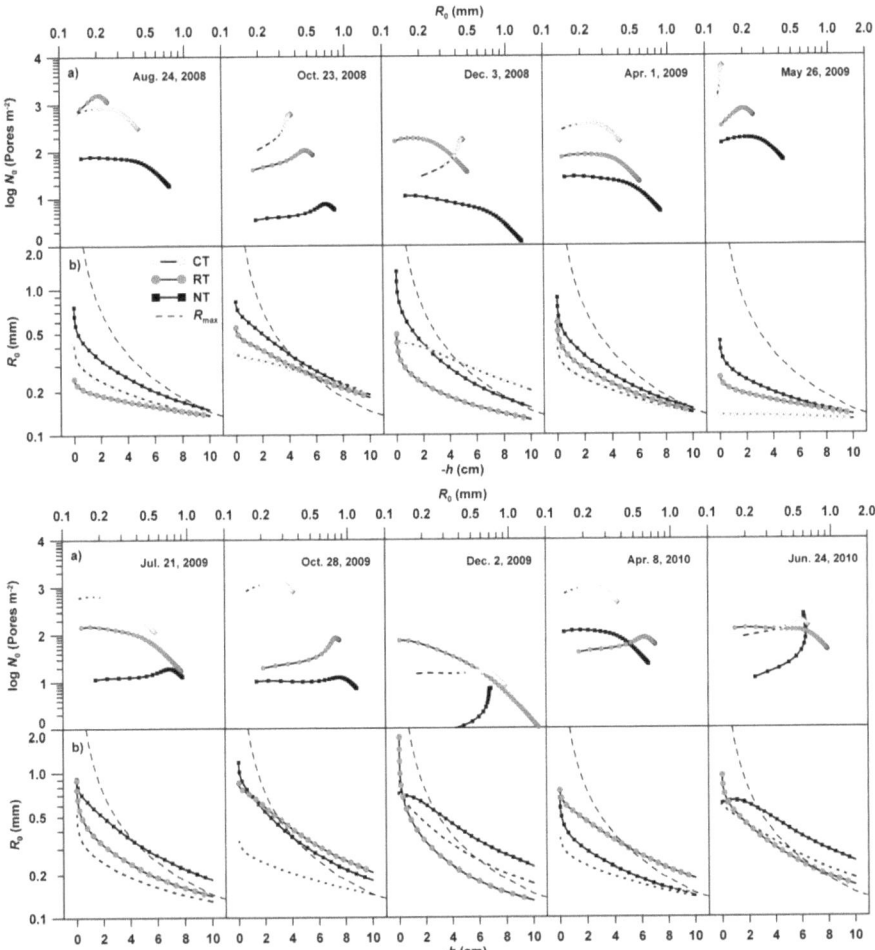

Figure 6-4. a) Number of flow-weighted mean pores N_0 versus flow-weighted mean pore radius R_0. The shaded areas denote the range of macropores (> 0.5 mm), the white area the range of large mesopores. b) Flow-weighted mean pore radius R_0 versus pressure head h in the range of the tension infiltrometer measurements. The dashed line denotes the maximum equivalent pore radius R_{max}. R_0 and N_0 were calculated using the approach of Reynolds et al. (1995). Representative mean curves for the treatments at each time of measurement are shown.

6.3.4 Water-Conducting Pore Characteristics

The derived conductivities were analyzed in terms of the water-conducting porosity (Figure 6-4), giving an indirect indicator of changes in the soil's pore structure. Generally, R_0 indicates that with increasing h, larger pores become water-conducting (Reynolds et al., 1995). This increase was largest for NT at most measurement times. NT also resulted in the greatest R_0 values over the measured range of h, whereas no distinct difference could be observed between CT and RT. This agrees with Moret and Arrue (2007a).

Compared to the maximum equivalent pore radius R_{max}, R_0 was smaller for most measurements. As stated by Reynolds et al. (1995), R_0, compared to storage-based R_{max}, better reflects the effects of pore restrictions, such as entrapped air bubbles or small unwetted zones. Thus, differences between R_{max} and R_0 indicate a reduction of the pore connectivity, leaving a certain fraction of porespace disconnected from the water-conducting pores. This difference was largest close to saturation, where flow is controlled by macropores. The connectivity was larger in the mespore range, agreeing with findings of Ehlers et al. (1995). The smallest difference over the measured pressure head range was observed for NT, followed by RT for most measurements. This indicates a good connectivity of the pores in the non-tilled soil, possibly due to good soil structure, a well established biological activity (e.g. earthworm burrows), and the presence of root channels. Recently, the relationship between K_s and ε_{macro} was also used as an indicator of the pore connectivity and tortuosity (Bodhinayake and Si, 2004; Luo et al., 2010). The regression slope between these quantities was much greater for NT (172.57) than for CT (57.99) and RT (56.50; Figure 6-5). This supports the assumption that the pores in the non-tilled soil were less tortuous and more effective in conducting water than in tilled soil. However, we acknowledge that observations of the pore connectivity and tortuosity were derived indirectly from the infiltration experiments and not measured directly (e.g. by tomographical image analysis).

The low R_0 values for CT in October 2009 and 2010 indicate that the pore-space connectivity was artificially disturbed due to previous ploughing. In May 2009, low R_0 values were measured for all treatments. This might be due to large unwetted zones and entrapped air, as the infiltration was measured after a long period without precipitation resulting in relatively low soil moisture (θ_i was 0.13, 0.14, and 0.21 m^3 m^{-3} for CT, RT, and NT, respectively, Table 6-1).

The largest R_0 that becomes water conducting at $h = 0$ (referred to as $R_{0,max}$) at most measurement times was found under NT (Figure 6-4a). Here, the largest water-conducting pore size was in the macropore-range (> 0.5 mm) with $R_{0,max}$ values of 1.33 and 1.18 mm in December 2008 and October 2009, respectively, and an overall mean of 0.79 mm. These pores dominate water infiltration under near-saturated conditions in this treatment, although N_0 was small (Messing and Jarvis, 1993; Moret and Arrue, 2007a).

RT gave a slightly smaller overall mean of $R_{0,max}$ (0.62 mm), but showed the single largest $R_{0,max}$ value (1.73 mm) in December 2009. Under CT, the flow-weighted pores had the smallest radii and were in the mesopore-range at most measurement times (overall mean of $R_{0,max}$: 0.42 mm). Generally, smaller R_0 values were compensated by a relatively greater number of these pore sizes (N_0, Figure 6-4a). Thus, for CT flow occurred in the smallest pore-size range, but with the greatest amount of these pores. This finding is supported by the hydraulically effective porosities ε (Figure 6-6). Over the whole analyzed time series, treatment means of ε_{meso} were in the order CT > RT > NT. As the differences were significant ($p < 0.05$) at most measured dates, differences in ε_{macro} among treatments where not systematically and hardly significant ($p < 0.05$).

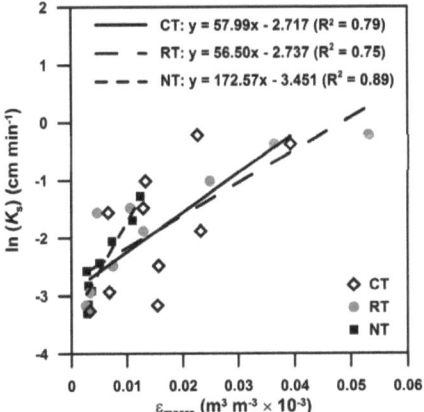

Figure 6-5. Relationship between the natural logarithm of the saturated hydraulic conductivity K_s and the water-conducting macroporosity ε_{macro}. Linear regressions were fitted to the treatment means of all measurement times. Larger slopes indicate a greater connectivity and smaller tortuosity of the water-conducting macropores (Luo et al., 2010).

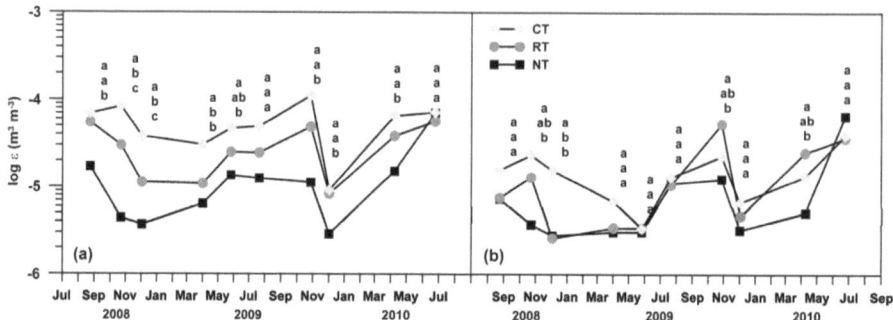

Figure 6-6. Temporal dynamics of hydraulically effective meso- (a) and macropores (b) ε according to the method of Bodhinayake et al. (2004) using the pore classification of Luxmoore (1981) with $r > 0.5$ mm for macropores and 5×10^{-3} mm $< r < 0.5$ mm for mesopores. Representative mean values for the tillage treatments (CT, RT and NT) are shown. Different letters indicate significant differences among treatments ($p < 0.05$).

6.3.5 Temporal vs. Management-Induced Dynamics of Hydraulic Properties

To differentiate between the impact of soil tillage and time of measurement, a multivariate ANOVA was applied (Table 6-3). A highly significant effect ($p < 0.01$) of tillage was found for K_{10} and ε_{meso}, with mean squares of the effect (MS) of 21.44 and 13.02, respectively. Both quantities describe water-conductance in the mesopores. Quantities related to water-conductance through macropores were less influenced by tillage, with a MS of 1.65 and 3.78 for K_s and ε_{macro}, respectively. The variability in water flow through macropores was more impacted by the measurement time. The interaction of both had a highly significant effect ($p < 0.01$) on K_{10} and ε_{meso}, but not on K_s and ε_{macro}. Thus, the variability of flow through mesopores could be explained sufficiently by an interaction of tillage and measurement time, whereas the flow through macropores could not. As discussed above, a high spatial variability was inherent in the abundance of macropores, which might exceed the impact of soil tillage and time. The high natural spatial variability might also explain most of the variability of α_{VG} and n.

The importance of time was also found in other studies (Alletto and Coquet, 2009; Bodner et al., 2008; Bormann and Klaassen, 2008; Hu et al., 2009). In both observed years, the hydraulically effective pores (ε_{macro} and ε_{meso}) strongly decreased after tillage during the winter periods, which might be a response to rainfall-induced pore sealing (Cameira et al., 2003; Mubarak et al., 2009; Figure 6-1 and 6-6). The decrease is followed by a gradually increase in spring and summer, possibly induced by biological activity, root development, and wetting / drying cycles, as proposed by Mubarak et al. (2009). This temporal dynamic should be considered when modeling soil water flow. Future investigations should focus on the parameterization of soil water models that account for temporal (e.g. post-tillage or seasonal) changes in the hydraulic properties and their underlying pore size development (e.g. Leij et al., 2002; Or et al., 2000; Schwärzel et al., 2010).

Table 6-3. Analysis of variance of soil hydraulic properties. The effects of soil tillage and measurement time on the hydraulic conductivity at saturation (K_s) and at $h = -10$ cm (K_{10}), the parameters of the van Genuchten/Mualem retention model (α_{VG} and n), and on the sums of water-conducting meso- and macropores (ε_{meso} and ε_{macro}) are listed.

	Tillage		Time		Tillage * Time	
	MS[a]	p[b]	MS	p	MS	p
K_s (cm s^{-1})	1.65	0.08*	6.27	<0.01***	0.71	0.35
K_{10} (cm s^{-1})	21.44	<0.01***	5.69	<0.01***	0.47	<0.01***
α_{VG} (m^{-1})	0.96	0.03**	0.67	0.02**	0.30	0.35
n (-)	0.19	0.05**	0.12	0.05**	0.12	0.03**
ε_{meso} (m^3 m^{-3})	13.02	<0.01***	4.63	<0.01***	1.09	<0.01***
ε_{macro} (m^3 m^{-3})	3.78	0.01**	7.59	<0.01***	1.04	0.19

[a] MS is the mean square of the effect.
[b] Values of p are indicated using three significance levels (*** $p<0.01$, ** $p<0.05$, * $p<0.1$).

6.4 Conclusion

We analyzed tension infiltrometer measurements over two consecutive years with respect to the hydraulic properties and the underlying water-conducting porosity. Our results revealed that hydraulic properties change dynamically in response to both tillage and natural controlling factors. Despite small temporal variations, NT resulted in the greatest values of θ_i and ρ_b, which might be due to natural compaction. $K(h)$ values were in the order CT > RT > NT, with the largest differences at $h = -10$ cm, where water flow is dominated by mesopores. A high spatial and temporal variability in the abundance of macropores was found under NT. Inverse estimation of the VGM model parameters showed only small temporal or tillage-induced changes for n, whereas α_{VG} was in the order CT < RT < NT, with high temporal variations under CT and RT. Regarding the water-conducting porosity,

NT resulted in the greatest R_0 values, whereas no distinct differences were observed between CT and RT. Smaller pore radii under RT and CT were compensated by a higher proportion of these pores. The results give indirect evidence that NT leads to a higher connectivity and smaller tortuosity of macropores, which might be due to a better established soil structure and biological activity. NT also resulted in a better temporal stability of both the pore network and the hydraulic properties.

A multivariate ANOVA revealed that variations in the mesopore-related quantities K_{10} and ε_{meso} could be explained by an interaction of tillage and time. By contrast, due to a high spatial variability, macropore-related quantities (K_s and ε_{macro}) could not be explained by those influence factors. The results give evidence that temporal changes of hydraulic properties (α_{VG}, K_s) should be considered when modeling soil water flow.

6.5 References

Alletto, L., and Coquet, Y., 2009. Temporal and spatial variability of soil bulk density and near-saturated hydraulic conductivity under two contrasted tillage management systems. Geoderma 152, 85-94.

Angulo-Jaramillo, R., Vandervaere, J.P., Roulier, S., Thony, J.L., Gaudet, J.P., and M. Vauclin, 2000. Field measurement of soil surface hydraulic properties by disc and ring infiltrometers - A review and recent developments. Soil & Tillage Research 55, 1-29.

Angulo-Jaramillo, R., Moreno, F., Clothier, B.E., Thony, J.L., Vachaud, G., FernandezBoy, E., and Cayuela, J.A., 1997. Seasonal variation of hydraulic properties of soils measured using a tension disk infiltrometer. Soil Sci. Soc. Am. J. 61, 27-32.

Ankeny, M.D., Kaspar, T.C., and Horton, R., 1988. Design for an automated tension infiltrometer. Soil Sci. Soc. Am. J. 52, 893-896.

Ankeny, M.D., Ahmed, M., Kaspar, T.C., and Horton, R., 1991. Simple field method for determining unsaturated hydraulic conductivity. Soil Sci. Soc. Am. J. 55, 67-470.

Bodhinayake, W., and Si, B.C., 2004. Near-saturated surface soil hydraulic properties under different land uses in the St Denis National Wildlife Area, Saskatchewan, Canada. Hydrological Processes 18, 2835-2850.

Bodhinayake, W., Si, B.C., and Xiao, C.J., 2004. New method for determining water-conducting macro- and mesoporosity from tension infiltrometer. Soil Sci. Soc. Am. J. 68, 760-769.

Bodner, G., Loiskandl, W., Buchan, G., and Kaul, H.-P., 2008. Natural and management-induced dynamics of hydraulic conductivity along a cover-cropped field slope. Geoderma 146, 317-325.

Bormann, H., and Klaassen, K., 2008. Seasonal and land use dependent variability of soil hydraulic and soil hydrological properties of two Northern German soils. Geoderma 145, 295-302.

Buczko, U., Bens, O., and Huttl, R.E., 2006. Tillage effects on hydraulic properties and macroporosity in silty and sandy soils. Soil Sci. Soc. Am. J. 70, 1998-2007.

Cameira, M.R., Fernando, R.M., and Pereira, L.S., 2003. Soil macropore dynamics affected by tillage and irrigation for a silty loam alluvial soil in southern Portugal. Soil & Tillage Research 70, 131-140.

Carey, S.K., Quinton, W.L., and Goellcr, N.T., 2007. Field and laboratory estimates of pore size properties and hydraulic characteristics for subarctic organic soils. Hydrological Processes 21, 2560-2571.

Daraghmeh, O.A., Jensen, J.R., and Petersen, C.T., 2008. Near-Saturated Hydraulic Properties in the Surface Layer of a Sandy Loam Soil under Conventional and Reduced Tillage. Soil Sci. Soc. Am. J. 72, 1728-1737.

Deurer, M., 2000. The dynamics of water and solute flow in the unsaturated zone of a coniferous forest: Measurement and numerical simulation Leibniz University of Hannover, Hannover, Germany.

Ehlers, W., Wendroth, O., and de Mol, F., 1995. Characterizing pore organization by soil physical parameters, in: Hartge, K.H. and Stewart, B.A. (Eds.), Soil Structure. Its Development and Function. Adv. Soil Sci., Lewis Publ., Boca Raton, pp. 257-275.

FAO, 1990. Guidelines for Soil Description. 3rd ed. FAO/ISRIC, Rome.

Gardner, W.R., 1958. Some steady-state solutions of the unsaturated moisture flow equation with application to evaporation from a water table. Soil Science 85, 228-231.

Hu, W., Shao, M.G., Wang, Q.J., Fan, J., and Horton, R., 2009. Temporal changes of soil hydraulic properties under different land uses. Geoderma 149, 355-366.

IUSS, 2007. World Reference Base for Soil Resources FAO, Rome.

Jarvis, N.J., and Messing, I., 1995. Near-saturated hydraulic conductivity in soils of contrasting texture measured by tension infiltrometers. Soil Sci. Soc. Am. J. 59, 27-34.

Lazarovitch, N., Ben-Gal, A., Šimůnek, J., and Shani, U., 2007. Uniqueness of soil hydraulic parameters determined by a combined wooding inverse approach. Soil Sci. Soc. Am. J. 71, 860-865.

Leij, F.J., Ghezzehei, T.A., and Or, D., 2002. Modeling the dynamics of the pore-size distribution. Soil & Tillage Research 64, 61-78.

Luo, L.F., Lin, H., and Schmidt, J., 2010. Quantitative Relationships between Soil Macropore Characteristics and Preferential Flow and Transport. Soil Sci. Soc. Am. J. 74, 1929-1937.

Luxmoore, R.J., 1981. Microporosity, Mesoporosity, and Macroporosity of Soil. Soil Sci. Soc. Am. J. 45, 671-672.

Marquardt, D.W., 1963. An algorithm for least-squares estimation of nonlinear parameters. Journal of the Society for Industrial and Applied Mathematics 11, 431-441.

Mesquita, M., Moraes, S.O., and Corrente, J.E., 2002. More adequate probability distributions to represent the saturated soil hydraulic conductivity. Scientia Agricola 59, 789-793.

Messing, I., and Jarvis, N.J., 1993. Temporal variation in the hydraulic conductivity of a tilled clay soil as measured by tension infiltrometers. Journal of Soil Science 44, 11-24.

Moret, D., and Arrue, J.L., 2007a. Dynamics of soil hydraulic properties during fallow as affected by tillage. Soil & Tillage Research 96, 103-113.

Moret, D., and Arrue, J.L., 2007b. Characterizing soil water-conducting macro- and mesoporosity as influenced by tillage using tension of infiltrometry. Soil Sci. Soc. Am. J. 71, 500-506.

Mubarak, I., Mailhol, J.C., Angulo-Jaramillo, R., Ruelle, P., Boivin, P., and Khaledian, M., 2009. Temporal variability in soil hydraulic properties under drip irrigation. Geoderma 150, 158-165.

Ndiaye, B., Molenat, J., Hallaire, V., Gascuel, C., and Hamon, Y., 2007. Effects of agricultural practices on hydraulic properties and water movement in soils in Brittany (France). Soil & Tillage Research 93, 251-263.

Nelson, D.W., and Sommers, L.E., 1982. Total carbon, organic carbon, and organic matter, in: P. et al. (Eds.), Methods of soil analysis Part 2: Chemical and microbiological properties. Agron. Monogr. ASA and SSSA, Madison, WI, pp. 539-579.

Or, D., Leij, F.J., Snyder, V., and Ghezzehei, T.A., 2000. Stochastic model for posttillage soil pore space evolution. Water Resources Research 36(7), 1641-1652.

Parkin, T.B., Meisinger, J.J., Chester, S.T., Starr, J.L., and Robinson, J.A., 1988. Evaluation of statistical estimation methods for lognormally distributed variables. Soil Sci. Soc. Am. J. 52, 323-329.

Ramos, T.B., Goncalves, M.C., Martins, J.C., van Genuchten, M.T., and Pires, F.P., 2006. Estimation of soil hydraulic properties from numerical inversion of tension disk infiltrometer data. Vadose Zone Journal 5, 684-696.

Reynolds, W.D., and Elrick, D.E., 1991. Determination of hydraulic conductivity using a tension infiltrometer. Soil Sci. Soc. Am. J. 55, 633-639.

Reynolds, W.D., Gregorich, E.G., and Curnoe, W.E., 1995. Characterization of water transmission properties in tilled and untilled soil using tension infiltrometers. Soil & Tillage Research 33, 117-131.

Reynolds, W.D., Bowman, B.T., Brunke, R.R., Drury, C.F., and Tan, C.S., 2000. Comparison of tension infiltrometer, pressure infiltrometer, and soil core estimates of saturated hydraulic conductivity. Soil Sci. Soc. Am. J. 64, 478-484.

Sauer, T.J., Clothier, B.E., and Daniel, T.C., 1990. Surface measurements of the hydraulic properties of a tilled and untilled soil. Soil & Tillage Research 15, 359-369.

Schaap, M.G., Leij, F.J., and van Genuchten, M.T., 2001. ROSETTA: a computer program for estimating soil hydraulic parameters with hierarchical pedotransfer functions. Journal of Hydrology 251, 163-176.

Schaap, M.G., and Leij, F.J., 2000. Improved Prediction of Unsaturated Hydraulic Conductivity with the Mualem-van Genuchten Model. Soil Sci. Soc. Am. J. 64(3), 843-851.

Schwärzel, K., Carrick, S., Wahren, A., Feger, K.-H., Bodner, G., and Buchan, G.D., 2011. Soil hydraulic properties of recently tilled soil under cropping rotation compared with 2-years-pasture: Measurement and modelling the soil structure dynamics. Vadose Zone Journal 10(1), 354-366.

Schwartz, R.C., and Evett, S.R., 2002. Estimating hydraulic properties of a fine-textured soil using a disc infiltrometer. Soil Sci. Soc. Am. J. 66, 1409-1423.

Schwen, A., Hernandez-Ramirez, G., Lawrence-Smith, E.J., Sinton, S.M., Carrick, S., Clothier, B.E., Buchan, G.D., and Loiskandl, W., 2011b. Hydraulic Properties and the Water-Conducting Porosity as Affected by Subsurface Compaction Using Tension Infiltrometers. Soil Sci. Soc. Am. J. 75(3), 822-831 (chapter 5 of this thesis).

Šimůnek, J., van Genuchten, M.T., and Sejna, M., 2006. The HYDRUS Software Package for Simulating the Two- and Three-Dimensional Movement of Water, Heat, and Multiple Solutes in Variably-Saturated Media. Technical Manual PC Progress, Prague, Czech Republik.

Šimůnek, J., Angulo-Jaramillo, R., Schaap, M.G., Vandervaere, J.P., and van Genuchten, M.T., 1998. Using an inverse method to estimate the hydraulic properties of crusted soils from tension-disc infiltrometer data. Geoderma 86, 61-81.

Šimůnek, J. and van Genuchten, M.T., 1997. Estimating unsaturated soil hydraulic properties from multiple tension disc infiltrometer data. Soil Science 162(6), 383-398.

Šimůnek, J. and van Genuchten, M.T., 1996. Estimating unsaturated soil hydraulic properties from tension disc infiltrometer data by numerical inversion. Water Resources Research 32(9), 2683-2696.

Soil Survey Staff, 2010. Keys to Soil Taxonomy. 11th ed. USDA-Natural Resources Conservation Service, Washington, DC.

Soracco, C.G., Lozano, L.A., Sarli, G.O., Gelati, P.R., and Filgueira, R.R., 2010. Anisotropy of Saturated Hydraulic Conductivity in a soil under conservation and no-till treatments. Soil & Tillage Research 109, 18-22.

Strudley, M.W., Green, T.R., and Ascough, J.C., 2008. Tillage effects on soil hydraulic properties in space and time: State of the science. Soil & Tillage Research 99, 4-48.

Van Genuchten, M.T., 1980. A closed-form equation for predicting the hydraulic conductivity of unsaturated soils. Soil Sci. Soc. Am. J. 44, 892-898.

Vereecken, H., Kasteel, R., Vanderborght, J., and Harter, T., 2007. Upscaling hydraulic properties and soil water flow processes in heterogeneous soils: A review. Vadose Zone Journal 6, 1-28.

Vomocil, J.A., 1965. Porosity, in: Black, C. A. (Eds.), Methods of soil analysis Part 1, Agron. Monogr. Ser. 9. ASA, Madison, WI, pp. 299-314.

Warrick, A.W., 1992. Models for disk infiltrometers. Water Resources Research 28, 1319-1327.

Warrick, A.W., and Nielsen, D.R., 1980. Spatial variability of soil physical properties in the field, in: Hillel, D. (Eds.), Applications of soil physics. Academic Press, Toronto, Canada.

Watson, K.W., and Luxmoore, R.J., 1986. Estimating macroporosity in a forest watershed by use of a tension infiltrometer. Soil Sci. Soc. Am. J. 50, 578-582.

Wooding, R.A., 1968. Steady infiltration from a shallow circular pond. Water Resources Research 4, 1259-1273.

Yoon, Y., Kim, J.G., and Hyun, S., 2007. Estimating soil water retention in a selected range of soil pores using tension disc infiltrometer data. Soil & Tillage Research 97, 107-116.

Zhou, X., Lin, H.S., and White, E.A., 2008. Surface soil hydraulic properties in four soil series under different land uses and their temporal changes. Catena 73, 180-188.

7. Time-variable soil hydraulic properties in near-surface soil water simulations for different tillage methods

Abstract

Simulating near-surface soil water dynamics is challenging since this soil compartment is temporally highly dynamic as response to climate and crop growth. For accurate simulations the soil hydraulic properties have to be proberly known. Although there is evidence that these properties are subject to temporal changes, they are set constant over time in most simulations studys. The objective of this study was to improve near-surface soil water simulations by accounting for time-variable hydraulic properties. Repeated tension infiltrometer measurements over two consecutive seasons were used to inversely estimate the hydraulic properties of a silt loam soil under different tillage – conventional (CT), reduced (RT), and no-tillage (NT). Simulated water dynamics with constant and time-variable hydraulic parameters were compared to observed data in terms of the soil water content and water storage in the near-surface soil profile (0-30 cm). The measurements indicate a considerable temporal variability in the saturated hydraulic conductivity, the field-saturated water content and the parameter α of the van Genuchten/Mualem model. Temporal variability was largest for CT and RT, whereas under NT, replicates of measured water contents and hydraulic properties showed a considerable large spatial variability. Simulations with time-constant hydraulic parameters led to underestimations of soil water dynamics in winter and early spring and overestimations during late spring and summer. The use of time-variable hydraulic parameters significantly improved simulation performance for all treatments, resulting in average relative errors below 13 %. Since simulation results agreed with observed water dynamics in two seasons, the applicability of inversely estimated hydraulic properties for soil water simulations is demonstrated. Thus, simulations that address applied questions in agricultural water management may be improved by using time-variable hydraulic parameters. The simulated water balance indicated that RT and NT result in better water storage than CT and therefore may increase water efficiency under water-limited climatic conditions.

7.1 Introduction

For many applied questions in the fields of crop production and agronomy, soil water dynamics are of fundamental importance. Modeling can be a valuable tool to optimize its management (Roger-Estrade et al. 2009). Such soil water modeling requires an accurate description of soil hydraulic properties, i.e. the soil water retention function $\theta(h)$ and the hydraulic conductivity function $K(h)$. Generally, these constitutive functions are assumed to be unchanged over time in most simulation studies. However, there is extensive empirical evidence that soil hydraulic properties are subject to temporal changes particularly in the near-saturated range where soil structure essentially influences water flow characteristics (Daraghmeh et al., 2008; Or et al., 2000). Especially, the structure of soil top layers is subject to changes during time, caused by wetting / drying cycles, biological activity, and agricultural operations (Leij et al., 2002; Mubarak et al., 2009). Soil tillage and management affect the hydraulic properties with consequences for the storage and movement of water, nutrients and pollutants, and for plant growth (Strudley et al., 2008). There is evidence, that soil tillage changes the soil pore-size distribution and the saturated hydraulic conductivity K_s (Mubarak et al., 2009; Or et al., 2000; Xu and Mermoud, 2003). The temporal variability of hydraulic properties can even exceed differences induced by crops, tillage, or land use (Alletto and Coquet, 2009; Bodner et al., 2008; Bormann and Klaassen, 2008; Hu et al., 2009; Schwen et al., 2011a; Zhou et al., 2008).

Compared to deeper soil layers, soil moisture close to the surface (0-30 cm) is subject to rapid changes as response to rainfall, infiltration, evaporation, and root water-uptake. Despite its importance for the supply of water and nutrients for crops, simulation of this highly dynamic soil compartment is difficult and requires adequate sets of hydraulic parameters (Šimůnek et al., 2003). For modeling nutrient or contaminant transport and for the assessment of different tillage methods, soil water simulations should be introduced that account for time-variable hydraulic properties (Mubarak et al., 2009; Or et al., 2000). However, only few studies have adressed this task. For instance, Or et al. (2000) introduced a model that describes temporal changes of the soil retention properties after tillage based on the pore-size distribution. Recently, Schwärzel et al. (2011) used the model to describe landuse-induced changes of the retention properties. However, the model has not been applied to a more complex time series of measured hydraulic parameters. Another approach was presented by Xu and Mermoud (2003), who used a time-dependent empirical decay function for K_s in soil water simulations.

To capture temporal variations in the soil hydraulic properties, suitable measurement methods have to be applied. As most of the temporal changes are expected to occur in the structural pores, due to changes in different groups of macro- and mesopores, field methods are preferable to laboratory methods (Angulo-Jaramillo et al., 2000; Hu et al., 2009; Yoon et al., 2007). To determine the near-saturated hydraulic properties of agricultural soils directly in the field, tension infiltrometry has become a commonly used method (Angulo-Jaramillo et al., 1997; Messing and Jarvis, 1993; Reynolds et al., 1995). The measurement of infiltration rates at pre-set supply pressure heads can be used to obtain the $K(h)$ function using Wooding's analytical equation (Reynolds and Elrick, 1991; Wooding, 1968) and to estimate inversely the parameters of a soil-water retention model (Angulo-Jaramillo et al., 2000; Schwartz and Evett, 2002; Šimůnek and van Genuchten, 1997; Šimůnek and van Genuchten, 1996). Although many authors have shown the reliability of inversely estimated retention parameters by comparison with direct measurement techniques, hardly any study assessed the applicability of the derived parameters in soil water simulations.

The main hypothesis underlying the presented study was that soil water simulations can be improved by accounting for temporal changes of near-surface soil hydraulic properties. Thus, the objectives of this study were 1) to capture temporal changes of soil hydraulic properties by repeated tension infiltrometer measurements, 2) to implement time-variable soil hydraulic properties in a soil water simulation. This study also aims to assess the feasibility of inversely estimated hydraulic parameters in soil water simulations. In the present paper, the effect of temporally changing hydraulic parameters in soil water simulation was tested and evaluated against measured data for different tillage methods. The simulation was used to assess the water efficiency of the different tillage methods afterwards.

7.2 Materials and Methods

7.2.1 Experimental Site

Measurements were obtained on an arable field near Raasdorf, Lower Austria (48°14'N 16°35'E). Climatic data were recorded at the site since 1998 by an automated weather station (Adcon Telemetry GmbH, Austria) (Figure 7-1). On the basis of the climatic data, the reference evapotranspiration ET_0 was calculated using the Penman-Monteith equation. The site has a mean annual precipitation R of 546 mm, a mean temperature of 9.8 °C, a mean relative humidity of 75 %, and an annual reference evapotranspiration ET_0 of 912 mm. The aridity index according to Middleton and Thomas (1992) is 0.60. Thus, the climate can be classified as dry subhumid (Middleton and Thomas, 1992) or as Cfb (maritime temperate climate with uniform precipitation distribution) according to Köppen (McKnight and Hess, 2000).

A field trial was established in 1997 to assess the following soil cultivation techniques: 1) Conventional tillage (CT) with moldboard ploughing and seedbed preparation using a harrow prior crop seeding; 2) Reduced tillage (RT) with a chisel plough to 10 cm for seedbed preparation; and 3) No-tillage (NT) using direct seeding technique. The three treatments were arranged in a randomized complete block design with three replicate plots. The soil can be classified as Chernozem in the WRB (IUSS, 2007) or as Typic Vermudoll in the US Soil Taxonomy (Soil Survey Staff, 2010). The humous A-horizon (0-30 cm) is followed by a AC-horizon (30-60 cm) over the mature silty sediments (C-horizon, >60 cm). Due to particle size analysis (Table 7-1) the texture throughout the profile can be classified as silt loam according to the FAO classification (FAO, 1990). The organic carbon content was 24 g kg^{-1} in the topsoil. In the two seasons analyzed for this study (2008/09 and 2009/10) the site was cropped with winter wheat (*Triticum aestivum* L.) in mid October and harvested in mid July of the following year (Figure 7-1).

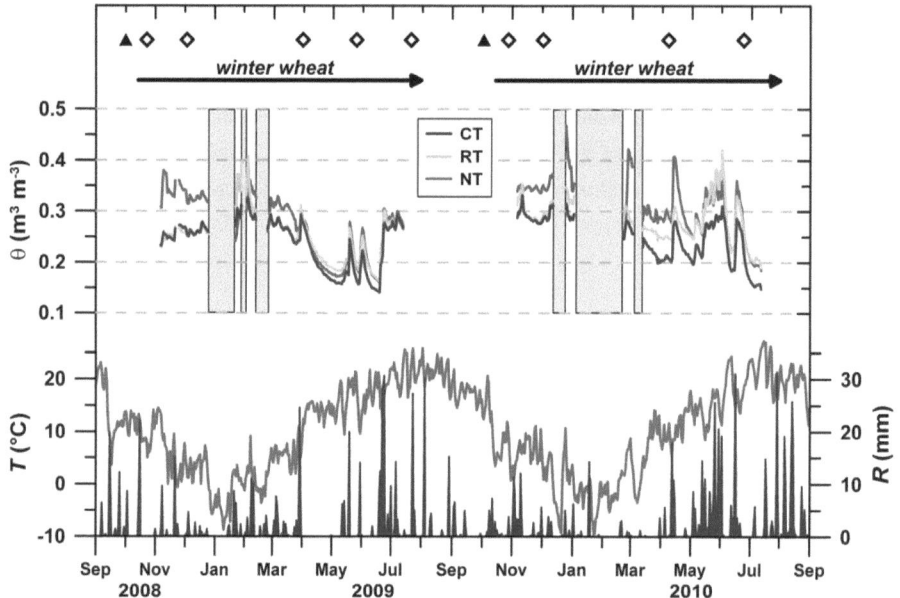

Figure 7-1. Soil water regime, climatic conditions, and times of measurements: Volumetric water content θ, measured in 10 cm depth for the different tillage treatments (mean of three replicate sensors), air temperature T (grey line); rainfall R (peaks), and times of infiltration measurements (diamonds), soil tillage (triangle), and crop growing period. The blanked areas indicate frost periods.

Soil water content in the field was continuously measured over two consecutive seasons using capacitance moisture sensors (C-Probe, Adcon Telemetry GmbH, Austria). For each treatment, three replicate probes were installed in depths of 10, 20, 40, 60, and 90 cm. The measurement interval was 15 min, and data was avaraged to daily values. For the comparison with simulation results in this study, only data of the shallow probes (10, 20, and 40 cm) were considered. The probes were installed after seeding in November and removed from the field prior harvest in July in both analyzed seasons. The water storage in the soil profile to a depth of 0.30 m S was derived from the water content measurement.

7.2.2 Sampling and Infiltration Measurements

Infiltration measurements used for this study were made nine times starting directly before crop seeding in October 2008 until shortly before harvest in July 2010, using three tension infiltrometers (Soil Measurement Systems Inc., Tucson, AZ) of the design described by Ankeny et al. (1988). The infiltrometer disc (diameter 0.20 m) was separate from the supply and tension control tubes. Before each measurement, the soil surface was prepared by carefully removing mulch and any above-ground plant material. We placed a nylon mesh on the smoothed soil to protect any macropores and used a 2 mm layer of quartz sand (diameter: 0.08-0.2 mm) between disc and soil to ensure good hydraulic contact. The supply pressure heads were -10, -4, -1, and 0 cm: the first two were maintained for approximately 50-60 min, and the last two for about 10-20 min. Preliminary tests found these durations to be sufficient to achieve steady-state infiltration. The water level in the water supply tube was observed visually at intervals of 20 s during the first 5 min after application of a supply pressure, and then in increasing intervals of 2-10 min. For each treatment, three replicate infiltration measurements were carried out. All experiments were conducted in the non-trafficked interrow area of a plot.

Before each infiltration measurement, soil samples were taken with steel cores near the measurement location to obtain the initial water content θ_i [$L^3 L^{-3}$]. Immediately after each measurement, another core sample was collected directly below the infiltration disc to quantify the final water content θ_f [$L^3 L^{-3}$], bulk density ρ_b [$M L^{-3}$], and total porosity φ [$L^3 L^{-3}$].

One soil profile per treatment was excavated and sampled with soil cores in depths of 5 cm, 40 cm, and 70 cm with three replicates during July 2009. The hydraulic properties of the subsoil layers were determined using pressure plate extractors (Soil Moisture Inc., USA) at $h = 0.2, 0.5, 1.0, 2.0$ and 3.0 bar. The RETC code (van Genuchten et al., 1991) was used to fit the parameters of the van Genuchten/Mualem model (referred to as VGM, van Genuchten, 1980; Table 7-1).

Table 7-1. Physical soil properties and hydraulic parameters at the experimental site. Texture (content of sand, silt and clay), bulk density ρ_b, saturated water content θ_s, saturated hydraulic conductivity K_s, and the VGM model parameters α_{VG} and n are listed.

Depth	Tillage	Sand	Silt	Clay	ρ_b	θ_s	K_s	α_{VG}	n
cm		——— kg kg^{-1} ———			g cm^{-3}	m^3 m^{-3}	m d^{-1}	m^{-1}	-
0-30		0.27	0.54	0.20	1.38	0.48			
30-60	CT	0.35	0.47	0.18	1.27	0.52	0.40	6.77	1.168
60-90		0.31	0.57	0.12	1.36	0.49	0.31	1.15	1.571
0-30		0.27	0.54	0.20	1.38	0.48			
30-60	RT	0.35	0.47	0.18	1.23	0.54	0.30	5.57	1.234
60-90		0.31	0.57	0.12	1.38	0.48	0.31	1.55	1.452
0-30		0.27	0.54	0.20	1.42	0.47			
30-60	NT	0.35	0.47	0.18	1.27	0.52	0.30	12.34	1.169
60-90		0.31	0.57	0.12	1.39	0.48	0.31	1.40	1.575

7.2.3 Near-Saturated Hydraulic Conductivity

We used the procedure described by Reynolds and Elrick (1991) and Ankeny et al. (1991) to determine $K(h)$ from the infiltration measurements. The variably saturated water flow equation can be analytically approximated for infiltration from a circular source with a constant pressure head at the soil surface, and with $K(h)$ described by Gardner's exponential model (1958):

$$K(h) = K_s \exp(\alpha_{Ga} h) \tag{7.1}$$

Here h is the pressure head [L], K_s is the saturated hydraulic conductivity [L T^{-1}], and α_{Ga} is the sorptive number. The analytical approximation derived by Wooding (1968) is:

$$q(h) = \left(\pi r_d^2 + \frac{4 r_d}{\alpha_{Ga}} \right) K(h) \tag{7.2}$$

where q is the steady-state infiltration rate [L^3 T^{-1}], and r_d is the radius of the disc [L]. Since Wooding's solution has two unknown variables, $K(h_0)$ and α_{Ga}, two steady-state fluxes at different tensions are required. Reynolds and Elrick (1991) among others derived $K(h)$ in the middle of an interval between two applied pressure heads h_i and h_{i+1}, assuming α_{Ga} to be constant over this interval. Assuming that Eq. [7.1] and [7.2] can be applied piecewise, then

$$\alpha_{Ga(i+1/2)} = \frac{\ln(q_i/q_{i+1})}{h_i - h_{i+1}}; \quad i = 1,\ldots,n-1 \qquad [7.3]$$

where n is the number of applied supply tensions. Using Eq. [7.2] gives:

$$K_{i+1/2} = \frac{q_i/q_{i+1}}{1 + [4/\pi r_d \alpha_{Ga(i+1/2)}]}; \quad i = 1,\ldots,n-1 \qquad [7.4]$$

K_s can be calculated from Eq. [7.1] using known values of $h_{i+1/2}$, $K_{i+1/2}$ and $\alpha_{Ga(i+1/2)}$ as follows:

$$K_s = \frac{K_{i+1/2}}{\exp(\alpha_{Ga(i+12)} h_{i+1/2})} \qquad [7.5]$$

7.2.4 Inverse Estimation of Soil Hydraulic Parameters

The inverse analysis of tension infiltrometer data requires numerical solution of the following modified Richards' equation for radially symmetric Darcian flow (Warrick, 1992):

$$\frac{\partial \theta}{\partial t} = \frac{1}{r}\frac{\partial}{\partial r}\left(rK\frac{\partial h}{\partial r}\right) + \frac{\partial}{\partial z}\left(K\frac{\partial h}{\partial z}\right) + \frac{\partial K}{\partial z} \qquad [7.6]$$

Here θ is the volumetric water content [$L^3 L^{-3}$], t the time [T], r is the radial coordinate [L] and z is the vertical coordinate [L], being positive upward with $z = 0$ on the soil surface. The initial and boundary conditions were defined as proposed by Šimůnek et al. (1998). To describe the unsaturated soil hydraulic properties, we used the VGM model. The soil water retention $S_e(h)$ and hydraulic conductivity $K(\theta)$ functions are given by:

$$S_e(h) = \frac{\theta(h) - \theta_r}{\theta_s - \theta_r} = \frac{1}{\left(1 + |\alpha_{vG} h|^n\right)^m} \qquad [7.7]$$

$$K(\theta) = K_s S_e^l \left[1 - \left(1 - S_e^{1/m}\right)^m\right]^2 \qquad [7.8]$$

Here S_e is the effective water content [-], θ_r and θ_s denote the residual and saturated water contents, respectively [L³ L⁻³], l is a pore-connectivity parameter [-], and α_{VG} [L⁻¹], n and m (= 1 – 1/n) are empirical parameters.

We formulated the objective function (OF) that is minimized during parameter estimation by combining the cumulative infiltration data, the $K(h)$ values calculated by Wooding's analysis, and θ_f. OF minimization was accomplished using the Levenberg-Marquardt nonlinear minimization method (Marquardt, 1963), as provided by the program HYDRUS 2D/3D (Šimůnek et al., 2006). For numerical solution of Eq. [7.6], a quasi-3D (axisymmetric) model geometry was chosen, as described by Šimůnek et al. (1998).

Initial values for the parameters were derived from the soil's texture using the Rosetta pedotransfer package Rosetta (Schaap et al., 2001; input parameters: soil texture and ρ_b). To reduce the amount of unknown variables, for all parameter estimations l was set constant at 0.5 (Ramos et al., 2006), and θ_r was fixed at 0.065 m³ m⁻³, as predicted by Rosetta (Lazarovitch et al., 2007). K_s was set to the value obtained by Wooding's analysis (Lazarovitch et al., 2007; Schwen et al., 2011b; Yoon et al., 2007). The remaining parameters θ_s, α_{VG}, and n were inversely estimated.

As the $\theta(h)$ and $K(h)$ relationships are highly nonlinear (Vereecken et al., 2007), representative mean parameters for each cultivation treatment were derived using the scaling approach. For each replicate measurement, data derived from the inverse parameter estimation and the data obtained by Wooding's equation were used. Following the approach of Schwärzel et al. (2010) and Vereecken et al. (2007), a conventional scaling procedure was applied in which scaling factors were estimated by minimizing the residual sum of square differences between the data and the scaled $K(h)$ and $\theta(h)$ reference curves. Representative means were calculated for every measurement time and as an overall treatment mean (Table 7-2).

7.2.5 Simulation Model

The vertical 1D Richards' equation was solved numerically using the Earth Science Modul within Comsol Multiphysics (Comsol AB). According to the observed soil horizons for the different tillage treatments, the geometry was devided into three layers: the surface soil (A-horizon) was between 0 and 0.15 m for RT and NT, and 0 - 0.30 m for CT, the AC-horizon down to 0.60 m, and the subsoil (C-horizon) from 0.60 m to 1.00 m.

Potential evaporation E_{pot} and transpiration T_{pot} were derived using the FAO 56 dual crop coefficient method (Allen et al., 1998). We used tabulated values for the crop coefficient K_c ($K_{c\ ini}$ = 0.4, $K_{c\ mid}$ = 1.15, $K_{c\ end}$ = 0.25, Allen et al., 1998) and observed crop development stages. We acknowledge that we did not calculate E_{pot} and T_{pot} treatment-specific but assumed them to be the same for the different tillage treatments.

R and E_{pot}, the latter reduced by a h-dependent reduction function, were applied as upper boundary condition (Bodner et al., 2007). T_{pot} was implemented via a sink term using a growth function, a linear decreasing root distribution function (Prasad, 1988) and a h-dependent reduction function according to Feddes et al. (2001) and Wu et al. (1999). We acknowledge that ET_{pot} was assumed to be the same for the different tillage treatments. Since the groundwater table was well below 10 m throughout the observation period, the lower boundary was defined by an unit gradient condition.

Soil water dynamics were simulated with a daily temporal discretization for the wheat growing seasons (Oct. 2008 – July 2009 and Oct. 2009 – July 2010; Figure 7-1). Simulations were made with constant and time variable VGM parameters (α_{VG}, K_s, and θ_s) of the upper soil layer for all tillage methods (Table 7-2). The measured parameter values were connected using cubic splines to allow a continous description. Since the temporal variability was expected to be negligible in the lower soil horizons, the hydraulic parameters were set constant (Table 7-1).

To get a comparison between observed and predicted values, O_i and P_i, the average relative error (ARE) was calculated as follows (Popova and Pereira, 2011):

$$ARE = \frac{1}{N}\sum_{i=1}^{N}\left(\frac{|O_i - P_i|}{O_i}\right) \qquad [7.9]$$

Here, N is the number of observations. For further comparison between O_i and P_i, regressions of the form P_i = slope × O_i + intercept were performed (Moret et al., 2007) and the determination coefficient R^2 was calculated. Root mean square errors (RMSE) were also calculated by (Ji et al., 2009; Popova and Pereira, 2011):

$$RMSE = \left[\frac{1}{N}\sum_{1}^{N}(P_i - O_i)^2\right]^{1/2} \qquad [7.10]$$

To account for the spatial variability of the observed values, O_i used in the differences was defined as mean (O_i) ± standard deviation. After validation, the model was used to calculate the water balance for the near-surface soil layer for the analyzed tillage treatments.

7.3 Results and Discussion

7.3.1 Climatic conditions and Soil Water Content

For the present study two consecutive wheat growing seasons were analyzed. R between October 15, 2008 and July 13, 2009 was 398 mm, which is in line with the long-term average. Remarkably, there was hardly any rain during April and May 2009 (Figure 7-1). Between October 15, 2009 and July 12, 2010, the rainfall was 395 mm. Potential evapotranspiration Et_{pot} in the seasons 2008/09 and 2009/10 was 366 and 332 mm, respectively, which can be divided into E_{pot} (117 and 101 mm), and T_{pot} (249 and 231 mm). In the growing seasons 2008/09 and 2009/10 the climatic water balance resulted in a surplus of 32 mm and 63 mm, respectively.

In response to climatic conditions and crop growth, the soil water content varied over time (Figure 7-1). With high water contents in winter and early spring and dryer periods between April and June 2009 the first analyzed year shows a broad variety of soil moisture conditions. Since there was sufficient rainfall in spring 2010, the soil moisture in the season 2009/10 was more balanced. During frost periods in winter, the shallow sensors did not give meaningful values that had to be excluded from further analysis (Figure 7-1). Only small differences in the near-surface water content occurred between treatments CT and RT, whereas NT resulted in significant higher soil moisture contents. This agrees with findings of Moret et al. (2007) and Moreno et al. (1997). However, as Moreno et al. (1997) found increasing treatment-induced differences with decreasing soil moisture, our results indicate larger differences at greater soil moisture and receded differences during dryer conditions (spring 2009 and summer 2010).

The variability within a treatment indicated by the replicate probes was considerable high (Figure 7-2). Since technical problems with the moisture sensors could be ruled out due to careful data processing, this variablility might be attributed to the natural spatial variability in the hydraulic properties (Schwen et al., 2011a). As the variability was smallest for CT and slightly larger for RT, it reflects the effect of spatial homogenization of repeated tillage operations. Contrarily, considerable differences among the replicate sensors under NT indicate a high spatial variability, especially in the season 2008/09. Under NT, the variability was much smaller in the season 2009/10 than in 2008/09. This may be explained by the fact that the probes were removed before harvest in July and reinstalled at slightly different positions after seeding in October 2009.

Table 7-2. Hydraulic parameters of the near-surface soil as derived from tension infiltrometer measurements. Scaled representative mean values of the saturated hydraulic conductivity K_s, the VGM model parameters $α_{VG}$ and n, and the soil water content at field-saturation $θ_s$ are listed.

	Tillage	Oct. 2008	Dec. 2008	Apr. 2009	May 2009	Jul. 2009	Oct. 2009	Dec. 2009	Apr. 2010	Jun. 2010	Overall Mean
K_s	CT	3.14	2.15	1.29	0.60	3.15	3.96	1.69	2.44	7.39	2.43
m d^{-1}	RT	2.20	0.61	0.76	0.55	3.29	4.04	0.53	1.26	9.43	1.79
	NT	0.78	1.10	0.86	0.62	2.65	11.57	3.04	5.23	9.90	0.96
$α$	CT	17.5	21.7	13.0	6.7	13.9	12.7	31.2	12.0	26.0	27.3
m^{-1}	RT	22.4	12.4	16.7	9.4	21.9	38.0	28.5	29.1	29.4	32.9
	NT	33.0	34.4	23.6	12.7	34.7	42.1	35.6	17.5	32.5	42.1
n	CT	1.809	1.813	1.519	1.858	1.413	1.582	1.549	1.531	1.661	1.812
-	RT	1.675	1.417	1.464	1.642	1.401	1.748	1.269	1.642	1.509	1.858
	NT	1.657	1.418	1.441	1.484	1.621	1.584	1.858	1.434	1.962	1.748
$θ_s$	CT	0.49	0.47	0.41	0.35	0.38	0.43	0.40	0.41	0.38	0.41
m^3 m^{-3}	RT	0.43	0.42	0.40	0.29	0.39	0.44	0.48	0.47	0.37	0.40
	NT	0.41	0.44	0.39	0.29	0.39	0.38	0.34	0.45	0.39	0.39

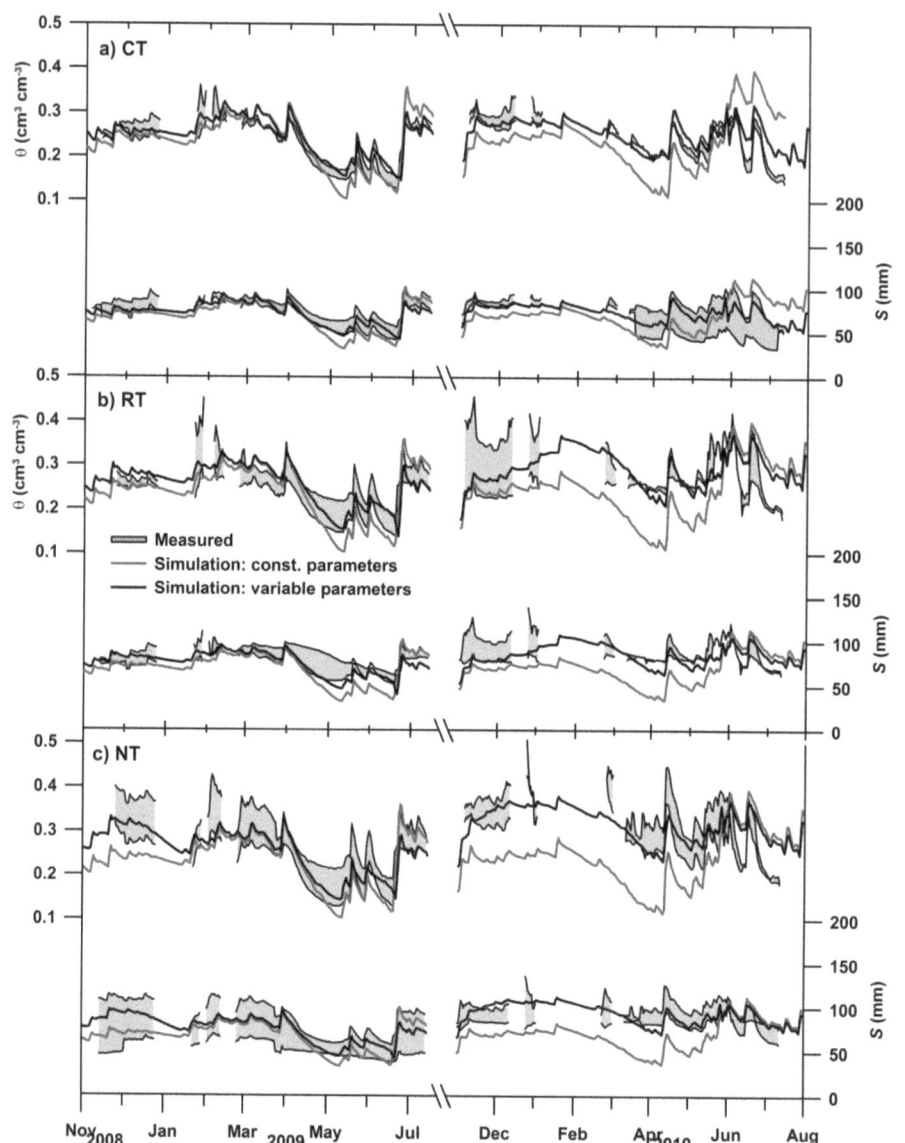

Figure 7-2. Simulated vs. measured volumetric soil water content θ in a depth of 10 cm (upper plots) and water storage S in the near-surface soil profile (0-30 cm; lower plots). θ and S (mean ± standard deviation) were measured using three replicate sensors per treatment in depths of 10, 20, and 40 cm (grey areas). The gaps in winter are due to frost periods (Figure 7-1). The results of simulations using constant and time-variable hydraulic properties are shown for the tillage treatments CT (a), RT (b), and NT (c).

7.3.2 Temporal Dynamics of Soil Hydraulic Properties

The temporal variability of soil hydraulic properties and its underlying water-conducting porosity at the experimental site has been discussed in detail by Schwen et al. (2011a; chapter 6 of this thesis). In both analyzed seasons, K_s and θ_s under CT and RT strongly decreased after tillage during winter (Table 7-2, Figure 7-1), which might be due to rainfall-induced pore sealing and settling (Cameira et al., 2003; Mubarak et al., 2009; Xu and Mermoud, 2003). The decrease is followed by a gradually increase in spring and summer, possibly induced by biological activity, root development, and wetting / drying cycles, as proposed by Mubarak et al. (2009). Beside considerable higher K_s values in October 2009 and June 2010, NT showed only small temporal variability and no systematic dynamic. The high K_s values in the non-tilled soil may be due to the existence of preferential flow paths (earthworm burrows), as reported by Moreno et al. (1997).

Regarding the VGM model parameters, n showed only small temporal dynamic for all treatments. However, α_{VG} showed a considerable temporal dynamic (Table 7-2). This dynamic was smallest for NT, indicating the relatively greater temporal stability of the hydraulic properties under this treatment (Schwen et al., 2011a). We found no systematic trend that reasonably explains the temporal dynamic of α_{VG} for both analyzed seasons. Possibly, this indicates that temporal dynamics of the VGM parameters are quite complex, even for the tilled treatments (CT, RT). Therefore, we did not apply a pore-size evolution model (e.g. Or et al., 2000), but used cubic splines to continuously describe the hydraulic parameters.

7.3.3 Performance of Near-Surface Water Simulations

The present study focusses on the simulation of water movement and storage in the highly dynamic near-surface soil compartment (0-30 cm). Therefore, in the following only results for the uppermost soil layer are discussed. However we acknowledge that we found a good agreement between measured and simulated soil water contents in the deeper soil layers for all simulations. For quality assessment of the simulations, we compared simulated and measured values of θ_{10} and S using Eq. [7.9] and [7.10]. Overall, the simulations resulted in soil water dynamics that agreed with the measured range for all treatments (Figure 7-2). Since the hydraulic parameters that were used in the topsoil were estimated inversely from tension infiltrometer measurements, this demonstrates the general applicability of this method for soil water simulations.

However, the degree of agreement between simulated and measured soil water dynamics differed among the tillage treatments and between simulations with temporally constant and variable hydraulic parameters (Figure 7-3, Table 7-3). We observed that simulations with time-constant sets of hydraulic parameters tend to underestimate θ_{10} and S in winter and spring (November 2008 – March 2009 and November 2009 – April 2010), whereas they resulted in overestimations during late spring and summer (June and July 2009 and 2010). Contrarily, application of time-variable hydraulic parameters significantly increased the agreement of both θ_{10} and S for all tillage treatments in both seasons. This was also reported by Xu and Mermoud (2003). Compared to simulations with constant parameters, values of ARE for θ_{10} and S approximately halved and the RSME for S was reduced by up to 93 % (Table 7-3). RSME values of θ_{10} for CT, RT, and NT were 0.042, 0.062, and 0.074 $m^3 \, m^{-3}$, respectively (Table 7-3), and thus in the same range as reported by Moret et al. (2007). As the ARE values for θ_{10} and S were below 13 %, the simulation results were satisfactory (Ji et al., 2009).

Within simulations with time-variable hydraulic parameters, the best agreement with ARE and RSME values for θ_{10} of 0.09 and 0.042 $m^3 \, m^{-3}$, respectively, (Table 7-3) and a correlation of $R^2 = 0.81$ (Figure 7-3) was found for CT, followed by NT (ARE = 0.09, $R^2 = 0.78$). Simulations for RT resulted in slightly larger differences and showed the weakest correlation (R^2 for $\theta_{10} = 0.65$, Figure 7-3). For all treatments, the agreement was better in the season 2008/09 than in 2009/10.

These results show that the accuracy of simulations of the near-surface soil water dynamics can be improved substantially by applying time-variable hydraulic parameters. The presented approach might help to improve the quality of soil water simulations, not only for the assessment of different tillage methods, but also for other applied questions in agricultural water management. As this study focussed on the comparison of temporally constant versus variable hydraulic parameters, we used the same values of E_{pot} and T_{pot} in all simulations. The accuracy in predicting the near-surface water dynamics in terms of different tillage methods might be further improved when treatment-dependent values for crop transpiration (rooting depth, leaf area index) are used (Xu and Mermoud, 2003).

To capture temporal and management-induced dynamics in the near-surface soil hydraulic properties, adequate and expeditious methods have to be applied. As demonstrated by this study, repeated tension infiltrometer measurements followed by the described data analysis procedure can meet this requirement. However, to derive physically-based descriptions of the complex temporal and management-induced dynamics of soil hydraulic properties, more research is necessary. We suggest that future research should focus on measurements at different temporal scales that might be correlated to underlying changes in the soil's structure.

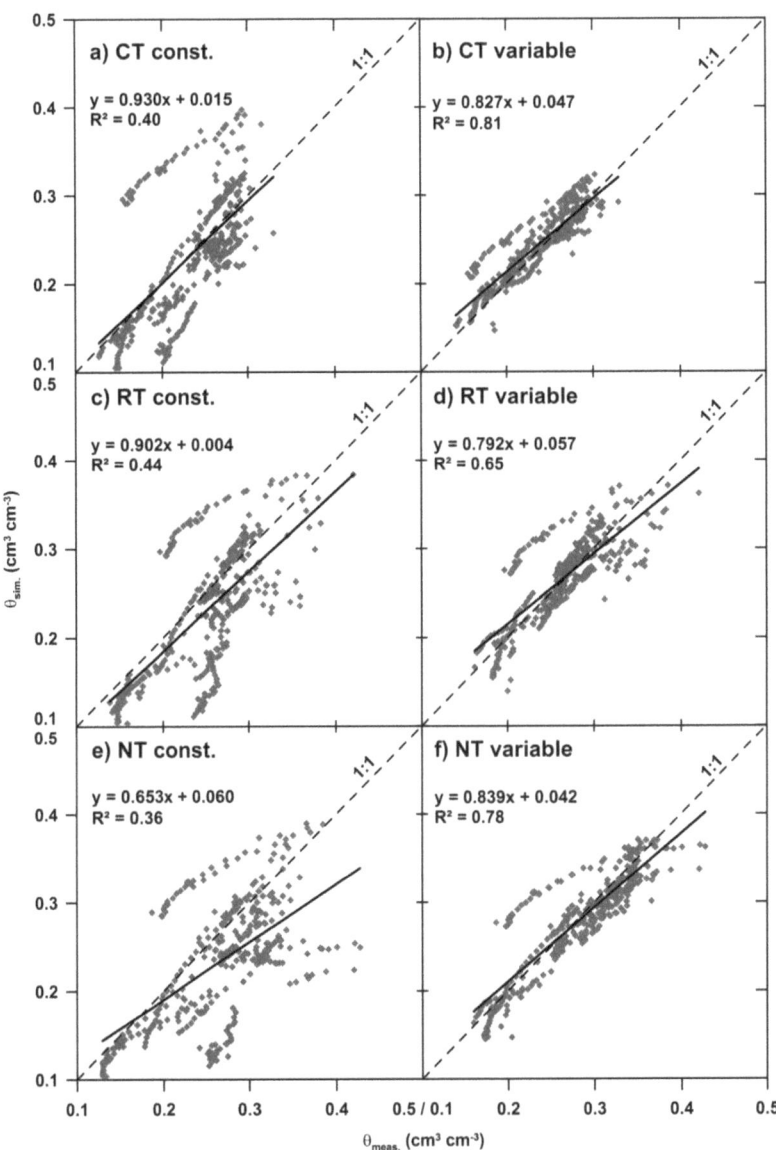

Figure 7-3. Correlation of simulated vs. measured soil volumetric water content θ in 10 cm depth. Results for simulations using constant (a, c, e) and time-variable hydraulic parameters (b, d, f) for each treatment are shown.

Table 7-3. Performance of the soil water simulation. Average relative errors ARE (Eq. 7.9) and root mean square errors RSME (Eq. 7.10) are listed for the volumetric water content in 10 cm depth, θ_{10}, and the water storage S in the near-surface soil profile (0-30 cm). Results for simulations with constant and time-variable hydraulic parameters are shown for all tillage treatments.

Tillage	Season	θ_{10}				S (0-30 cm)			
		ARE (-)		RSME ($m^3 m^{-3}$)		ARE (-)		RSME (mm)	
		const.	var.	const.	var.	const.	var.	const.	var.
CT	2008/2009	0.11	0.07	0.037	0.035	0.05	0.01	5.4	2.1
	2009/2010	0.32	0.11	0.080	0.049	0.16	0.02	14.3	2.8
	Overall	0.21	0.09	0.062	0.042	0.11	0.01	10.7	2.5
RT	2008/2009	0.13	0.08	0.065	0.072	0.09	0.03	10.0	4.0
	2009/2010	0.31	0.13	0.077	0.045	0.25	0.08	23.9	9.5
	Overall	0.21	0.10	0.072	0.062	0.17	0.05	18.1	7.2
NT	2008/2009	0.15	0.07	0.060	0.073	0.02	0.001	2.7	0.2
	2009/2010	0.35	0.12	0.101	0.073	0.20	0.03	22.3	3.9
	Overall	0.24	0.09	0.083	0.074	0.10	0.01	15.6	2.7

7.3.4 Water Balance Simulation

Applying the simulation with time-variable hydraulic parameters, the water balance of the soil profile was calculated for the different tillage treatments and both analyzed seasons (Table 4). To assess tillage-induced differences in the soil water storage S, the difference between the time of crop planting (mid October) and harvest (mid July) was calculated. There were considerable differences between the treatments in the change of stored water and percolation losses, whereas there were hardly any differences in E_{act} and T_{act}. CT resulted in the smallest increase in water storage in both seasons. This is mainly due to a high percolation below 100 cm and a slightly higher evaporation. The water storage in the soil profile at seeding in October 2009 was much smaller than in the previous year. This may be due to the dry climatic conditions in spring 2009, that have not been compensated at the time of seeding. In the growing season 2009/10, the replenishment of water stored in the soil profile is much more effective under treatments RT ($\Delta S = 59$ mm) and NT ($\Delta S = 53$ mm) than under CT ($\Delta S = 39$ mm). The latter treatment shows a much higher loss due to percolation (37 mm). The result suggests that RT and NT lead to an increased water storage and a smaller loss of water due to percolation, especially under dryer initial soil conditions. These cultivation practices may increase water efficiency by keeping it in the part of the soil that can be reached by plant roots. This finding agrees with Xu and Mermoud (2003) and Moreno et al. (1997), who reported that tillage-induced differences may become more evident under dryer climatic conditions. Thus, reduced or no-tillage practices may be used to mitigate draught stress under dry subhumid or semi-arid climatic conditions.

Table 7-4. Water balance for the two simulated wheat growing seasons. Rainfall R (measured), actual evaporation (E_{act}), actual transpiration (T_{act}), actual evapotranspiration (ET_{act}), the water storage in the soil profile S at crop plant and harvest and its change throughout the growing season ΔS, and the percolation below 100 cm P are listed for the analyzed tillage treatments.

Tillage	Season	R	E_{act}	T_{act}	ET_{act}	S Plant	S Harvest	ΔS	P
					mm				
CT	2008/2009	398.0	113.0	232.3	345.3	178.9	207.3	28.4	24.3
	2009/2010	395.4	93.1	226.9	320.0	156.6	195.3	38.7	36.7
RT	2008/2009	398.0	111.8	229.1	340.9	168.7	203.7	35.0	22.0
	2009/2010	395.4	92.2	226.6	318.8	151.5	210.5	59.0	17.6
NT	2008/2009	398.0	108.5	233.0	341.5	170.8	204.9	34.1	22.4
	2009/2010	395.4	93.5	228.0	321.5	167.2	220.2	53.0	20.9

7.4 Conclusion

The present study reveals the applicability of repeated tension infiltrometer measurements to capture temporal and tillage-induced changes in soil hydraulic properties. As classical simulations of the soil water use temporally constant hydraulic parameters, we used a Richards' equation water simulation that enables the flexible definition of these important control quantities. Simulated water content and storage in the near-surface soil compartments under different tillage treatments were compared to measured data. The performance of the simulation could be improved significantly using time-variable hydraulic parameters, regardless the tillage. By giving meaningful results, our simulations demonstrate the applicability of inversely estimated hydraulic parameters for soil water simulations. The simulated water balance indicated that RT and NT may increase water storage in the soil, especially under dryer climatic conditions.

7.5 References

Allen, R.G., Pereira, L.S., Raes, D. & Smith, M., 1998. Crop evapotranspiration – Guidelines for computing crop water requirements. FAO Irrigation and drainage paper 56, Food and Agriculture Organization of the United Nations, Rome.

Alletto, L., and Coquet, Y., 2009. Temporal and spatial variability of soil bulk density and near-saturated hydraulic conductivity under two contrasted tillage management systems. Geoderma 152, 85-94.

Angulo-Jaramillo, R., Vandervaere, J.P., Roulier, S., Thony, J.L., Gaudet, J.P., and M. Vauclin, 2000. Field measurement of soil surface hydraulic properties by disc and ring infiltrometers - A review and recent developments. Soil & Tillage Research 55, 1-29.

Angulo-Jaramillo, R., Moreno, F., Clothier, B.E., Thony, J.L., Vachaud, G., FernandezBoy, E., and Cayuela, J.A., 1997. Seasonal variation of hydraulic properties of soils measured using a tension disk infiltrometer. Soil Sci. Soc. Am. J. 61, 27-32.

Ankeny, M.D., Kaspar, T.C., and Horton, R., 1988. Design for an automated tension infiltrometer. Soil Sci. Soc. Am. J. 52, 893-896.

Ankeny, M.D., Ahmed, M., Kaspar, T.C., and Horton, R., 1991. Simple field method for determining unsaturated hydraulic conductivity. Soil Sci. Soc. Am. J. 55, 67-470.

Bodner, G., Loiskandl, W., Buchan, G., and Kaul, H.-P., 2008. Natural and management-induced dynamics of hydraulic conductivity along a cover-cropped field slope. Geoderma 146, 317-325.

Bodner, G., Loiskandl, W., and Kaul, H.-P., 2007. Cover crop evapotranspiration under semi-arid conditions using FAO dual crop coefficient method with water stress compensation. Agr. Water Management 93, 85-98.

Bormann, H., and Klaassen, K., 2008. Seasonal and land use dependent variability of soil hydraulic and soil hydrological properties of two Northern German soils. Geoderma 145, 295-302.

Cameira, M.R., Fernando, R.M., and Pereira, L.S., 2003. Soil macropore dynamics affected by tillage and irrigation for a silty loam alluvial soil in southern Portugal. Soil & Tillage Research 70, 131-140.

Carey, S.K., Quinton, W.L., and Goellcr, N.T., 2007. Field and laboratory estimates of pore size properties and hydraulic characteristics for subarctic organic soils. Hydrological Processes 21, 2560-2571.

Daraghmeh, O.A., Jensen, J.R., and Petersen, C.T., 2008. Near-Saturated Hydraulic Properties in the Surface Layer of a Sandy Loam Soil under Conventional and Reduced Tillage. Soil Sci. Soc. Am. J. 72, 1728-1737.

FAO, 1990. Guidelines for Soil Description. 3rd ed. FAO/ISRIC, Rome.

Feddes, R.A, Hoff, H., Bruen, M., Dawson, T., de Rosnay, P., Dirmeyer, O., Jackson, R.B., Kabat, P., Kleidon, A., Lilly, A. & Pitman, A.J., 2001. Modeling root water uptake in hydrological and climate models. Bulletin of the American Meteorological Society 82, 2797-2809.

Gardner, W.R., 1958. Some steady-state solutions of the unsaturated moisture flow equation with application to evaporation from a water table. Soil Science 85, 228-231.

Hu, W., Shao, M.G., Wang, Q.J., Fan, J., and Horton, R., 2009. Temporal changes of soil hydraulic properties under different land uses. Geoderma 149, 355-366.

IUSS, 2007. World Reference Base for Soil Resources FAO, Rome.

Ji, X.B., Kang, E.S., Zhao, W.Z., Zhang, Z.H., and Jin, B.W., 2009. Simulation of heat and water transfer in a surface irrigated, cropped sandy soil. Agr. Water Management 96, 1010-1020.

Lazarovitch, N., Ben-Gal, A., Šimůnek, J., and Shani, U., 2007. Uniqueness of soil hydraulic parameters determined by a combined wooding inverse approach. Soil Sci. Soc. Am. J. 71, 860-865.

Leij, F.J., Ghezzehei, T.A., and Or, D., 2002. Modeling the dynamics of the pore-size distribution. Soil & Tillage Research 64, 61-78.

Marquardt, D.W., 1963. An algorithm for least-squares estimation of nonlinear parameters. Journal of the Society for Industrial and Applied Mathematics 11, 431-441.

McKnight, T.L. and Hess, D., 2000. Climate Zones and Types: The Köppen System. *In*: Physical Geography: A Landscape Appreciation. Upper Saddle River, NJ: Prentice Hall. ISBN 0130202630.

Messing, I., and Jarvis, N.J., 1993. Temporal variation in the hydraulic conductivity of a tilled clay soil as measured by tension infiltrometers. Journal of Soil Science 44, 11-24.

Middleton, N. and Thomas, D.S.G., 1992. World Atlas of Desertification. First Edition, United Nations Environment Programme (UNEP), Edward Arnold, London, ISBN 340691662.

Moret, D., Braud, I., and Arrúe, J.L., 2007. Water balance simulation of a dryland soil during fallow under conventional and conservation tillage in semiarid Aragon, Northeast Spain. Soil & Tillage Research 92, 251-263.

Mubarak, I., Mailhol, J.C., Angulo-Jaramillo, R., Ruelle, P., Boivin, P., and Khaledian, M., 2009. Temporal variability in soil hydraulic properties under drip irrigation. Geoderma 150, 158-165.

Or, D., Leij, F.J., Snyder, V., and Ghezzehei, T.A., 2000. Stochastic model for posttillage soil pore space evolution. Water Resources Research 36(7), 1641-1652.

Prasad, R., 1988. A linear root water uptake model. J. Hydrol. 99, 297-306.

Popova, Z. and Pereira, L.S., 2011. Modelling for maize irrigation scheduling using long term experimental data from Plovdiv region, Bulgaria. Agricultural Water Management 98(4), 675-683.

Ramos, T.B., Goncalves, M.C., Martins, J.C., van Genuchten, M.T., and Pires, F.P., 2006. Estimation of soil hydraulic properties from numerical inversion of tension disk infiltrometer data. Vadose Zone Journal 5, 684-696.

Reynolds, W.D., and Elrick, D.E., 1991. Determination of hydraulic conductivity using a tension infiltrometer. Soil Sci. Soc. Am. J. 55, 633-639.

Reynolds, W.D., Gregorich, E.G., and Curnoe, W.E., 1995. Characterization of water transmission properties in tilled and untilled soil using tension infiltrometers. Soil & Tillage Research 33, 117-131.

Roger-Estrade, J., Richard, G., Dexter, A.R., Boizard, H., de Tourdonnet, S., Bertrand, M. & Caneill, J., 2009. Integration of soil structure variations with time and space into models for crop management. A review. Agron. Sustain. Dev. 29, 135-142.

Schaap, M.G., Leij, F.J., and van Genuchten, M.T., 2001. ROSETTA: a computer program for estimating soil hydraulic parameters with hierarchical pedotransfer functions. Journal of Hydrology 251, 163-176.

Schwärzel, K., Carrick, S., Wahren, A., Feger, K.-H., Bodner, G., and Buchan, G.D., 2011. Soil hydraulic properties of recently tilled soil under cropping rotation compared with 2-years-pasture: Measurement and modelling the soil structure dynamics. Vadose Zone Journal 10(1), 354-366.

Schwartz, R.C., and Evett, S.R., 2002. Estimating hydraulic properties of a fine-textured soil using a disc infiltrometer. Soil Sci. Soc. Am. J. 66, 1409-1423.

Schwen, A., Bodner, G., Scholl, P., Buchan, G.D., and Loiskandl, W., 2011a. Temporal dynamics of soil hydraulic properties and the water-conducting porosity under different tillage. Soil & Tillage Research 113(2), 89-98 (chapter 6 of this thesis).

Schwen, A., Hernandez-Ramirez, G., Lawrence-Smith, E.J., Sinton, S.M., Carrick, S., Clothier, B.E., Buchan, G.D., and Loiskandl, W., 2011b. Hydraulic Properties and the Water-Conducting Porosity as Affected by Subsurface Compaction Using Tension Infiltrometers. Soil Sci. Soc. Am. J. 75(3), 822-831 (chapter 5 of this thesis).

Šimůnek, J., van Genuchten, M.T., and Sejna, M., 2006. The HYDRUS Software Package for Simulating the Two- and Three-Dimensional Movement of Water, Heat, and Multiple Solutes in Variably-Saturated Media. Technical Manual PC Progress, Prague, Czech Republik.

Šimůnek, J., Jarvis, N.J., van Genuchten, M.T., and Gärdenäs, A., 2003. Review and comparison of models for describing non-equilibrium and preferential flow and transport in the vadose zone. J. Hydrol. 272, 14-35.

Šimůnek, J., Angulo-Jaramillo, R., Schaap, M.G., Vandervaere, J.P., and van Genuchten, M.T., 1998. Using an inverse method to estimate the hydraulic properties of crusted soils from tension-disc infiltrometer data. Geoderma 86, 61-81.

Šimůnek, J. and van Genuchten, M.T., 1997. Estimating unsaturated soil hydraulic properties from multiple tension disc infiltrometer data. Soil Science 162(6), 383-398.

Šimůnek, J. and van Genuchten, M.T., 1996. Estimating unsaturated soil hydraulic properties from tension disc infiltrometer data by numerical inversion. Water Resources Research 32(9), 2683-2696.

Soil Survey Staff, 2010. Keys to Soil Taxonomy. 11th ed. USDA-Natural Resources Conservation Service, Washington, DC.

Strudley, M.W., Green, T.R., and Ascough, J.C., 2008. Tillage effects on soil hydraulic properties in space and time: State of the science. Soil & Tillage Research 99, 4-48.

van Genuchten, M.T., 1980. A closed-form equation for predicting the hydraulic conductivity of unsaturated soils. Soil Sci. Soc. Am. J. 44, 892-898.

van Genuchten, M.T., Leij, F.J., and Yates, S.R., 1991. The RETC Code for Quantifying the Hydraulic Functions of Unsaturated Soils. Version 6.0, US Salinity Laboratory, USDA.

Vereecken, H., Kasteel, R., Vanderborght, J., and Harter, T., 2007. Upscaling hydraulic properties and soil water flow processes in heterogeneous soils: A review. Vadose Zone Journal 6, 1-28.

Warrick, A.W., 1992. Models for disk infiltrometers. Water Resources Research 28, 1319-1327.

Wooding, R.A., 1968. Steady infiltration from a shallow circular pond. Water Resources Research 4, 1259-1273.

Wu, J., Zhang, R., and Gui, S., 1999. Modeling soil water movement with water uptake by roots. Plant and Soil 215, 7-17.

Xu, D. and A. Mermoud, 2003. Modeling the soil water balance based on time-dependent hydraulic conductivity under different tillage practices. Agricultural Water Management 63, 139-151.

Yoon, Y., Kim, J.G., and Hyun, S., 2007. Estimating soil water retention in a selected range of soil pores using tension disc infiltrometer data. Soil & Tillage Research 97, 107-116.

Zhou, X., Lin, H.S., and White, E.A., 2008. Surface soil hydraulic properties in four soil series under different land uses and their temporal changes. Catena 73, 180-188.

8. Final conclusion

In the present thesis, the impacts of subsurface compaction and different tillage techniques on the soil hydraulic properties and the water-conducting pore characteristics were assessed. The application of different levels of subsurface compaction affected the hydraulic properties, reducing K_s, θ_m, R_0, and α_{VG}. A high susceptibility of the analyzed silt loam soil to the applied compaction was found, since K_s of the heavy compacted soil was 81% less than that of the loosened soil. As response to increasing compaction, the main water conductance was shifted from macropores into mesopores. This might be due to a distortion and reduction of the connectivity of the macropore-network. As a result, the easy-measurable K_s value may be used as a proxy for the hydraulical impact of soil compaction. As a consequence for crop growth and soil water modeling purposes, a reduced K_s in compacted subsurface layers should be incorporated.

Infiltration measurements under different tillage treatments revealed that soil hydraulic properties change dynamically over time as a response to both tillage and natural controlling factors. The near-saturated hydraulic conductivity $K(h)$ was in the order CT > RT > NT, with the largest differences where water flow is dominated by mesopores. The inverse parameter estimation of the VGM model revealed, that α_{VG} was in the order CT < RT < NT, with high temporal variations under CT and RT. Under NT, a high spatial variability of the macropore network dominates water flow. This treatment is temporally more stable and resulted in the best connectivity of the water-conducting pores. Smaller pore radii under RT and CT were compensated by a higher proportion of these pores. Variations in mesopore-related quantities could be explained by an interaction of tillage and time, whereas due to a high spatial variability, macropore-related quantities could not.

The results of chapter 5 and 6 show that soil hydraulic properties were altered differently by subsurface compaction and different tillage methods. As the applied subsurface compaction had a strong impact on the hydraulic conductivity, tillage-induced differences were smaller and overlain by temporal dynamics. The results indicate that tension infiltrometer measurements can be used to quantify tillage- and compaction-induced changes in the soil hydraulic properties and underlying pore characteristics. Thus, agricultural impacts on soil hydraulic properties and its spatial and temporal variability can be captured by this measurement technique and the applied subsequent data analysis. However, there is a need for a better differentiation between treatment-induced and natural spatial variability in soil hydraulic properties. Future investigations may use different field trial layouts to enable a geostatistical interpretation of measurements.

Another outcome of this thesis is that parameters of the VGM model that were inversely estimated from tension infiltrometer measurements could be used for soil water simulations.

Simulations of the near-surface soil water dynamics were improved significantly using time-variable hydraulic properties. The soil water balance simulation for the different analyzed tillage techniques revealed that RT and NT may increase water storage in the near-surface soil, especially under dryer climatic conditions.

9. References

Angulo-Jaramillo, R., Vandervaere, J.P., Roulier, S., Thony, J.L., Gaudet, J.P., and M. Vauclin, 2000. Field measurement of soil surface hydraulic properties by disc and ring infiltrometers - A review and recent developments. Soil & Tillage Research 55, 1-29.

Angulo-Jaramillo, R., Moreno, F., Clothier, B.E., Thony, J.L., Vachaud, G., Fernandez-Boy, E., and Cayuela, J.A., 1997. Seasonal variation of hydraulic properties of soils measured using a tension disk infiltrometer. Soil Sci. Soc. Am. J. 61, 27-32.

Daraghmeh, O.A., Jensen, J.R., and Petersen, C.T., 2008. Near-Saturated Hydraulic Properties in the Surface Layer of a Sandy Loam Soil under Conventional and Reduced Tillage. Soil Sci. Soc. Am. J. 72, 1728-1737.

Hu, W., Shao, M.G., Wang, Q.J., Fan, J., and Horton, R., 2009. Temporal changes of soil hydraulic properties under different land uses. Geoderma 149, 355-366.

Messing, I., and Jarvis, N.J., 1993. Temporal variation in the hydraulic conductivity of a tilled clay soil as measured by tension infiltrometers. Journal of Soil Science 44, 11-24.

Mubarak, I., Mailhol, J.C., Angulo-Jaramillo, R., Ruelle, P., Boivin, P., and Khaledian, M., 2009. Temporal variability in soil hydraulic properties under drip irrigation. Geoderma 150, 158-165.

Or, D., Leij, F.J., Snyder, V., and Ghezzehei, T.A., 2000. Stochastic model for posttillage soil pore space evolution. Water Resources Research 36(7), 1641-1652.

Reynolds, W.D., Gregorich, E.G., and Curnoe, W.E., 1995. Characterization of water transmission properties in tilled and untilled soil using tension infiltrometers. Soil & Tillage Research 33, 117-131.

Strudley, M.W., Green, T.R., and Ascough, J.C., 2008. Tillage effects on soil hydraulic properties in space and time: State of the science. Soil & Tillage Research 99, 4-48.

van Genuchten, M.T., 1980. A closed-form equation for predicting the hydraulic conductivity of unsaturated soils. Soil Sci. Soc. Am. J. 44, 892-898.

Xu, D. and A. Mermoud, 2003. Modeling the soil water balance based on time-dependent hydraulic conductivity under different tillage practices. Agricultural Water Management 63, 139-151.

Yoon, Y., Kim, J.G., and Hyun, S., 2007. Estimating soil water retention in a selected range of soil pores using tension disc infiltrometer data. Soil & Tillage Research 97, 107-116.

10. Index of tables

Table 4-1. List of scientific publications in journals that are listed in the Science Citation Index (SCI) 3

Table 4-2. List of presentations on international scientific conferences 3

Table 5-1. Unsaturated hydraulic conductivity $K(h)$ as obtained from Wooding's equation of the tension infiltrometer measurements 16

Table 5-2. Physical properties within the compacted soil layer and results of the infiltration measurements for different steps of subsoil compaction 21

Table 6-1. Bulk density ρ_b, total porosity φ and volumetric water content prior the infiltration measurement θ_i 38

Table 6-2. Results of the inverse estimation of the van Genuchten/Mualem soil water retention model 43

Table 6-3. Analysis of variance of soil hydraulic properties 49

Table 7-1. Physical soil properties and hydraulic parameters at the experimental site 62

Table 7-2. Hydraulic parameters of the near-surface soil as derived from tension infiltrometer measurements 67

Table 7-3. Performance of the soil water simulation 72

Table 7-4. Water balance for the two simulated wheat growing seasons 73

11. Index of figures

Figure 5-1. Plots of the subsurface compaction trial arranged in a latin square and locations of the infiltration measurements 8

Figure 5-2. Hydraulically effective macroporosity and mesoporosity 17

Figure 5-3. Water-conducting porosity θ_m and number of hydraulically effective pores N_m for different pore radii r_m 18

Figure 5-4. Flow-weighted mean pore radius R_0 versus pressure head h and number of flow-weighted mean pores N_0 versus flow-weighted mean pore radius R_0 20

Figure 6-1. Climatic conditions, soil cultivation and times of measurements 31

Figure 6-2. Near-saturated hydraulic conductivity $K(h)$ as a function of the supply pressure head h 41

Figure 6-3. Correlation between hydraulic conductivity at $h = -10$ cm (K_{10}) and at saturation (K_s) versus initial volumetric moisture content θ_i 44

Figure 6-4. Number of flow-weighted mean pores N_0 versus flow-weighted mean pore radius R_0 45

Figure 6-5. Relationship between the natural logarithm of the saturated hydraulic conductivity K_s and the water-conducting macroporosity ε_{macro} 47

Figure 6-6. Temporal dynamics of hydraulically effective meso- and macropores 48

Figure 7-1. Soil water regime, climatic conditions, and times of measurements 60

Figure 7-2. Simulated vs. measured volumetric soil water content θ in a depth of 10 cm and water storage S in the near-surface soil profile 68

Figure 7-3. Correlation of simulated vs. measured soil volumetric water content θ 71

i want morebooks!

Buy your books fast and straightforward online - at one of world's fastest growing online book stores! Environmentally sound due to Print-on-Demand technologies.

Buy your books online at
www.get-morebooks.com

Kaufen Sie Ihre Bücher schnell und unkompliziert online – auf einer der am schnellsten wachsenden Buchhandelsplattformen weltweit! Dank Print-On-Demand umwelt- und ressourcenschonend produziert.

Bücher schneller online kaufen
www.morebooks.de

VDM Verlagsservicegesellschaft mbH
Heinrich-Böcking-Str. 6-8   Telefon: +49 681 3720 174   info@vdm-vsg.de
D - 66121 Saarbrücken   Telefax: +49 681 3720 1749 www.vdm-vsg.de

Printed by Books on Demand GmbH, Norderstedt / Germany